Aspects of Quality Management in Value Creating in the Industry 5.0 Way

Industry 5.0 suggests a new stage of industrial growth that expands upon earlier stages of industrialization, emphasizing human-centered approaches to technology and digital sustainability. With its innovative approach, Industry 5.0 will contribute to the resolution of the manufacturing–social need mismatch issue. In contrast to other industrial revolutions that placed more emphasis on the financial aspects of sustainability, the Industry 5.0 vision places more emphasis on social demands and human centricity.

This book *Aspects of Quality Management in Value Creating in the Industry 5.0 Way* focuses on the challenges that companies in the field of quality management in Industry 5.0 face, particularly in relation to client value aspects. The book devotes a lot of space to the issues of client satisfaction, cybersecurity, e-commerce, TQM, and collaborative work between robots and humans in the company.

Features:

- Characterizes the new role of value for customer 5.0 in the augmented era
- Analyzes the collaborative work between robots and humans in Industry 5.0 conditions
- Investigates the complex relationship between satisfaction, awareness, perception, attitude, and demographics, as well as examining how technological advances and market performance impact client satisfaction

Includes:

- E-client in the cyber-security aspect
- Multi-Agent Technology (MAT) to maintain Total Quality Management (TQM) in manufacturing and MAT's role in TQM
- A novel structure for innovation, "Innovation Control (IC)," to integrate creative thinking and business strategy
- Industry 5.0 inside the automotive sector

- Technetronic Education (TE) in Industry 5.0: advantages, challenges, and implications
- Ethical aspects and challenges associated with developing technologies

This book *Aspects of Quality Management in Value Creating in the Industry 5.0 Way* serves as a future road map, guiding readers through the complexities of industrial progress.

Academic researchers, along with senior undergraduate and graduate students, are the primary target audience.

Aspects of Quality Management in Value Creating in the Industry 5.0 Way

Edited by Mohamed Abouhawwash,
Joanna Rosak-Szyrocka, and
Shashi Kant Gupta

CRC Press
Taylor & Francis Group
Boca Raton London New York

CRC Press is an imprint of the
Taylor & Francis Group, an **informa** business

Contents

Preface

A new range of very sophisticated technological solutions have been pushed toward industry in recent years by the emphasis on smart production systems. As a matter of fact, smart manufacturing systems often include smart quality management optimization skills to save costs and increase overall production efficiency via the use of technology-driven approaches like Industry 4.0. Nonetheless, Industry 5.0 suggests a new stage of industrial growth that expands upon earlier stages of industrialization, emphasizing human-centered approaches to technology and digital sustainability. Digital sustainability and Industry 5.0 are closely related. The incorporation of cutting-edge digital technology into production processes is a fundamental component of Industry 5.0. These technologies may be used to streamline industrial procedures, save energy and trash usage, and raise general productivity. Industry 5.0 signals the end of the world's industrial revolution. It seeks to achieve social objectives beyond employment and growth by putting the welfare of people at the heart of manufacturing systems, so providing strong prosperity for the sustainable development of all humankind. With its innovative approach, Industry 5.0 will contribute to the resolution of the manufacturing–social need mismatch issue. In contrast to other industrial revolutions that placed more emphasis on the financial aspects of sustainability, the Industry 5.0 vision places more emphasis on social demands and human centricity. A human-centric, sustainable, and resilient industry is what Industry 5.0 aims to achieve. It is powered by advanced technologies and is divided into three categories:

1. individualized human–machine interaction (artificial intelligence (AI), robotics, augmented and virtual reality, cyber-physical system (CPS), digital twins of products, processes, and entire systems);
2. manufacturing system simulation (controls-IoT-security (CIS), digital twins of products, processes, and entire systems); and
3. data transmission, storage, and analysis technologies (industrial internet of things (IIoT), cyber-security, big data analytics, and edge computing).

The book's primary goal is to examine the evolving role of value for customers in the context of Industry 5.0. It also examines the nature of the industry 5.0 revolution and how 21st-century customer requirements have evolved as a result of quality management considerations, intelligent systems, and satisfaction issues.

Contributors

Julee Banerji
IIMS, Sai Balaji Group of Institutions
Pune, Maharashtra, India

Shovona Choudhury
Amity University
Nayasarai, Ranchi, Jharkhand, India

Chandra Kumar Dixit
Dr. Shakuntala Misra National Rehabilitation University
Lucknow, Uttar Pradesh, India

Shashi Kant Gupta
Computer Science and Engineering Department, Eudoxia Research
    University
New Castle, DE, USA

V. Suresh Kumar
Vel Tech Multi Tech Dr. Rangarajan Dr. Sakunthala Engineering College
Avadi, Chennai, India

Gilbert C. Magulod Jr.
Cagayan State University
Tuguegarao City, Philippines

Anna-Marie Pelser
Faculty of Economic and Financial Sciences, North-West University
South Africa

Joanna Rosak-Szyrocka
Częstochowa University of Technology
Częstochowa, Poland

Zafarullah Sahito
Department of Education, Sukkur IBA University
Sukkur, Sindh, Pakistan

Faculty of Economic and Financial Sciences, North-West University
South Africa

Flavio Boccia
Department of Economic and Legal Studies,
University of Naples "Parthenope"
Italy

Neha Mishra
Chitkara Business School, Chitkara University, Punjab
India

Atul Garg
Chitkara University Institute of Engineering and Technology, Chitkara
    University, Punjab
India

Himanshu Lad
Faculty of School of Liberal Arts & Management Studies, P P Savani
    University
Surat, Gujarat, India

Aparna Vajpayee
Faculty of School of Liberal Arts & Management Studies, P P Savani
    University
Surat, Gujarat, India

Parag Sanghani
Faculty of School of Liberal Arts & Management Studies, P P Savani
    University
Surat, Gujarat, India

Hasim Deari
University of Tetova, Faculty of Economics,
North Macedonia

Eldian Balla
University "Aleksander Moisiu", Faculty of Business,
Albania

Vidhi Tyagi
Amity College of Commerce & Finance, Amity University
Noida, India

Shikha Mittal
Master of Business Administration, AKTU, RKGIT,
Ghaziabad, India

Abha Gupta
Department of Management Studies, RDIAS, Guru Gobind Singh
    Indraprastha University
India

Shubham Sharma
Department of Mechanical Engineering, National Institute of Technology
Hamirpur, Himachal Pradesh, India

Somesh Kumar Sharma
Department of Mechanical Engineering, National Institute of Technology
Hamirpur, Himanchal Pradesh, India

Priya V.
Department of Computer Science and Engineering, Dr N.G.P. Institute of
    Technology
Coimbatore, Tamil Nadu, India

Zamal Mohamed Zubair
Firebird Institute of Research in Management
Coimbatore, Tamil Nadu, India

Vipin C.
Department of Computer Science and Engineering, Vellore Institute of
    Technology
India

Pranav S.
Department of Artificial Intelligence and Data Science, Dr.N.G.P. Institute
    of Technology
Coimbatore, Tamil Nadu, India

Noor Ahmed
Department of Education, Sukkur IBA University
Sukkur Sindh, Pakistan

Insaf Ali
Department of Education, Sukkur IBA University
Sukkur Sindh, Pakistan

Rahul Joshi
Department of Journalism & Mass Communication, Manav Rachna
    International Institute of Research and studies,
Haryana, India

About the editors

Mohamed Abouhawwash received the B.Sc. and M.Sc. degrees in statistics and computer science from Mansoura University, Mansoura, Egypt, in 2005 and 2011, respectively, and the joint Ph.D. degree in statistics and computer science from Michigan State University, East Lansing, MI, USA, and Mansoura University, Egypt, in 2015. Currently, he holds significant academic positions at Distinguished Institutions, at Michigan State University, East Lansing, MI, USA. Additionally, he serves as an Associate Professor at the Department of Mathematics, Faculty of Science, Mansoura University, Egypt. He is dedicated to advancing knowledge that transcends geographical boundaries, as evidenced by his role as a Visiting Scholar at the Department of Mathematics and Statistics, Faculty of Science, Thompson Rivers University, Kamloops, BC, Canada, during 2018. He is a Distinguished Researcher and Academician, widely recognized for his outstanding contributions to the fields of computational intelligence, machine learning, and image reconstruction. With an illustrious career, he has published over 180 papers in esteemed journals, including notable publications like *IEEE Transactions on Evolutionary Computation*, *IEEE Transactions on Medical Imaging*, *IEEE Transactions on Emerging Topics in Computational Intelligence*, *Artificial Intelligence Review*, *Expert Systems with Applications*, *Swarm and Evolutionary Computation*, *Knowledge-Based Systems*, and *Applied Soft Computing*. In addition to his prolific research output, he has showcased his expertise by authoring several edited books published by reputable academic publishers such as Springer, Wiley, and Taylor & Francis. His impact on the academic community is further amplified through his editorial board service in numerous prestigious journals and conferences. Throughout his illustrious career, he has received recognition for his academic excellence, notably being honored with the best master's and Ph.D. Thesis Awards from Mansoura University in 2012 and 2018, respectively.

Joanna Rosak-Szyrocka, H-41 index Google Scholar, H-11 index Web of Science, H-12 index Scopus. She is Assistant Professor, Erasmus+ Coordinator at the Faculty of Management, Czestochowa University of Technology, Poland. She specialized in the fields of digitalization, Industry 5.0, quality 4.0, education, IoT, AI, and quality management. Participant in multiple Erasmus+ teacher mobility programs: Italy, UK, Slovenia, Hungary, Czech Republic, Slovakia, and Board of France. She cooperates with many universities both in the country and abroad. She is in the Editorial *PLoS One Journal, PeerJ Journal*. Advisory Board of *Heliyon Journal*. Associate Editor for Cogent Business and Management, Taylor & Francis. She is a Guest Editor of *Entertainment Computing Journal, Elsevier Journal, Resources MdPI, IJERPH MdPI, Energies MdPI, Sustainability MdPI, Springer Discover Sustainability Journal, Frontiers*, and *Elsevier Measurement Journal*. Reviewer for a number of prestigious journals like *IEEE, Elsevier, MdPI, Frontiers, Sage, Springer*, and *Emerald*. Member of the Research Team at the Faculty of Management of the Częstochowa University of Technology 2022 and the interdisciplinary team 2022. Member of the team for surveying deans' offices and the course of studies in the field of occupational health and safety and member of the technical team for evaluation at the Faculty of Management of the Częstochowa University of Technology. She was awarded a diploma of recognition from the Dean of the Faculty of Management of the Częstochowa University of Technology for her publishing activities, as well as a medal from the Dean of the Faculty of Management of the Częstochowa University of Technology in gratitude for long-term cooperation with the Faculty of Management (2017). In 2023, she was awarded the medal of the President of the Republic of Poland for long-term service.

Shashi Kant Gupta Post-Doctoral Fellow and Researcher, Computer Science and Engineering, Eudoxia Research University, USA in collaboration with Eudoxia Research Centre, India. ORCID: 0000-0001-6587-5607. He is working as an Honorary Senior Research Fellow, Department of Scientific Research, Innovation and Training of Scientific and Pedagogical Staff, University of Economics and Pedagogy, Karshi City, Uzbekistan. He is working as a Research Collaborator and Invited Visiting Senior Scientist at the Research Institute of IoT and Cybersecurity, Department of Electronic Engineering, National Kaohsiung University of Science and Technology, Taiwan. He has completed his Ph.D. (CSE) from Integral University, Lucknow, UP, India. He worked as Assistant Professor in the Department of Computer Science and Engineering, PSIT, Kanpur. He is also the founder and CEO of CREP Pvt Ltd, Lucknow, Uttar Pradesh, India. He is a member of Spectrum IEEE and other international organizations for research activities. He has published many research papers in reputed international journals,

SCOPUS, and SCI indexed journals. His research focuses on performance enhancement through Cloud computing, Big Data Analytics, IoT, and Computational Intelligence-Based Education. He is currently working as a reviewer in various international journals like *BJIT*. He has published many Indian patents in the field of information technology, computer science and management. One Indian patent is under grant approval. He has more than 12 years of teaching experience and 2 years of industrial experience.

Introduction

Industry 5.0 will enable a new era of collaborative production by fusing the precision, speed, and data processing of intelligent systems with the imagination, adaptability, and problem-solving skills of people. The book consists of 12 chapters. Chapter 1 presents the new role of value for customer 5.0 in the augmented era (Industry 5.0). The authors analyzed what the industry 5.0 revolution looked like and how the customer requirements of the 21st century have changed. Chapter 2 focuses on the concept pertaining to Industry 5.0, wherein the notion of collaborative work between robots and humans is put forth as an alternative to their conventional competition. Chapter 3 explores customer satisfaction in stock broking firms, investigating the complex relationship between satisfaction, awareness, perception, attitude, and demographics, as well as examining how technological advances and market performance impact satisfaction in the stock broking industry. Chapter 4 analyzes e-client safety as a determinant of value in the cyber-security aspect. Chapter search and find which relevant factors help companies establish the safety of e-clients in the cyber-security aspect. Chapter 5 investigates the latest paradigm of the industrial revolution, Industry 5.0, and its profound effect on human resource (HR) practices. Chapter 6 outlines the primary e-commerce concerns from the seller's and the buyer's perspectives. Chapter 7 explores Multi-Agent Technology (MAT) to maintain Total Quality Management (TQM) in manufacturing and analyzes MAT's role in TQM. Chapter 8 investigates the elements of upgrading human–machine coordinated effort with regard to Industry 5.0 inside the automotive sector. Chapter 9 presents a novel structure for innovation called "Innovation Control (IC)" to integrate creative thinking, business strategy, and the innovation environment to accomplish revenues and a competitive edge while making innovation easier to comprehend, implement, and integrate into routine activities. Chapter 10 provides scientific insights for quality education with the characteristics of educational leadership and management, that is, diversity, decentralization, problem-solving, resilience, sustainability, environmental harmony, and value creation. Chapter 11 investigates Technetronic Education (TE) in the context of Industry 5.0, aiming to understand its advantages, challenges, and implications. Chapter 12 examines the ethical aspects and challenges associated with developing technologies and proposes viable strategies to address their legal and regulatory concerns.

Chapter 1

New role of value for customer 5.0 in augmented era

Joanna Rosak-Szyrocka, Shashi Kant Gupta, and Flavio Boccia

> *Industry 5.0 will make the factory a place where creative people can come and work to create more personalized and human experiences for employees and their customers.*
>
> Esben Østergaard

1.1 INTRODUCTION

A new reality is emerging in and around Ukraine as a result of political, economic, social, and technical developments. It is defined by the characteristics of the VUCA environment. The idea was first put out by American military specialists to refer to a dynamic fighting environment, but since the turn of the twenty-first century, it has been extensively employed to characterize the contemporary corporate environment. Volatility, uncertainty, complexity, and ambiguity are represented by the first letters of the acronym VUCA. VUCA reality—in other words—means instability, uncertainty, complexity, and ambiguity of the environment for business operations. With the advent and quick expansion of the Internet and mobile computer communication devices, the information society entered a new phase of growth. A digital culture, defined by the digitization of all social activities, is starting to emerge on a new technical foundation. The world is evolving into an information society's digital ecology (Rosak-Szyrocka, Żywiołek & Shahbaz 2023).

The challenges of the twenty-first century are as follows: the economy is becoming more globalized; the Fourth Industrial Revolution is occurring; the economy and society are becoming more digitalized; the COVID-19 pandemic has created a complex epidemiological situation; the demographic gap is widening; business strategies and business models are changing; and modern organizations must radically restructure business practices to focus on core competencies while simultaneously moving everything that does not pertain to these competencies into so-called ecosystems, such as clouds, networks, and platforms (*Higher Education and Digitalization in Perspective of Use of Internet, Integration of Digital Technology, Digital*

DOI: 10.1201/9781032677040-1

Public ... 2023; Pan et al. 2023; Rosak-Szyrocka, Apostu & Akkaya 2023). When faced with global issues, an organization should have a flexible structure, be mobile, evolve dynamically, and be able to swiftly create and bring new goods to market (Mitrofanova, Mitrofanova & Margarov 2023). To meet the constantly shifting needs of the market, manufacturers compete. Production lines must thus be able to respond to new needs in a flexible, intelligent, and adaptable manner. Managers of manufacturing and business have come to the conclusion that business and industrial output should be integrated. Significant progress in industrial processes and strategies is necessary for such an integration. Furthermore, it can only be attained by combining different parts of an organization, such as suppliers, manufacturing processes, and clients (Nahavandi 2019). The industrial revolution is being facilitated by a number of enabling technologies known as Industry 5.0, such as mass customization, hyper-personalization, cobots, and digital twins. Every sector has been impacted by the digital revolution, including marketing and corporate organizations of all sizes. Marketing professionals have developed unique and inventive approaches by using technology advancements resulting from both industrial and social revolutions.

The aim of the chapter is to present the new role of value for customer 5.0 in the era of augmented era (Industry 5.0). There are studies on Industry 5.0 (Demir & Cicibaş 2018; Akundi et al. 2022; Alves, Lima & Gaspar 2023) and customer 5.0 (Carvalho & Alves 2023; Lee & Lee 2020), but they concern the challenges and characteristics of these issues. Our study is innovative because it analyzes how the requirements of the customer 5.0— the prosumer—have changed in the era of Industry 5.0 and how they understand value.

1.2 LITERATURE REVIEW

1.2.1 From Industry 1.0 to Industry 5.0

Industry 5.0 revolution is presented in Figure 1.1. Industry 1.0 (IR1.0) had its start in the early eighteenth century with the invention of steam and automation. Mechanization caused the spinning industry's output to expand eightfold during this time. Productivity developed and

improved in huge enterprises as a result of the primary component of this revolution—steam. In the spinning industry, steam power took the role of human muscular effort throughout this era. After steam power was introduced, advancements in steamships and steam locomotives led to yet another significant shift during this time because they slashed the amount of time it took for people and commodities to travel huge distances (Yavari & Pilevari 2020). Humans doing "manual and dexterous work" with some mechanical and manually operated machine tool aid are referred to as the Operator 1.0 generation (Mourtzis, Angelopoulos & Panopoulos 2022).

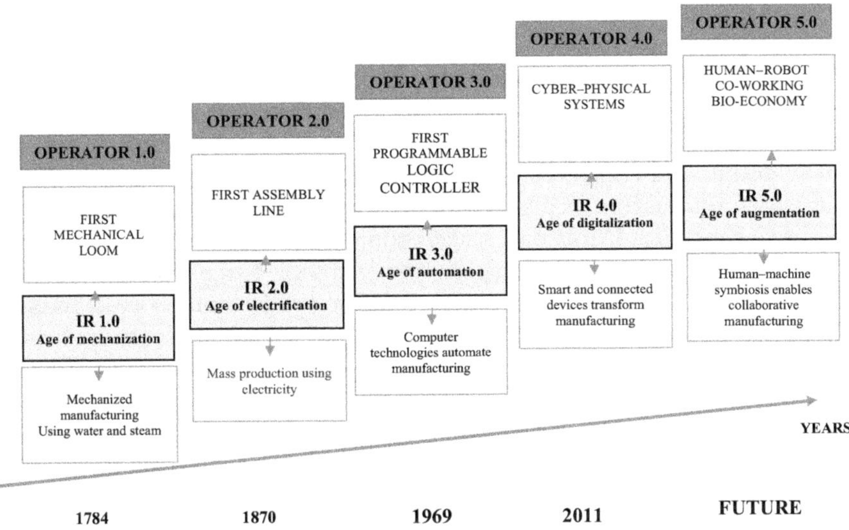

Figure 1.1 Industry 5.0 revolution.

Source: Own study based on: Demir, Döven & Sezen (2019).

The introduction of electricity (IR2.0) as the primary power source occurred at the start of the 1800s. The simplicity of use of electricity over steam and water was one of its benefits. This characteristic allowed this power supply to be used in various and unique equipment. During this time, management tools advanced significantly and enhanced industry performance and efficiency in addition to the introduction of electricity. During this time, division of labor was a common practice that raised profitability. The Operator 2.0 generation comprises human workers who use computer tools such as business information systems, collaborative networked communication (CNC) machine tools, and computer-aided everything (CAx) tools, as well as numerical control (NC) operating systems, to do "assisted work."

Industry 3.0 (IR3.0) saw the development of production automation systems in plants. These technologies make a variety of activities easier for staff to do and the idea of mass manufacturing less daunting. The Operator 3.0 generation is a representation of a person working in "collaborative work," or human–robot cooperation, with computers, other machines, and robots (Mourtzis, Angelopoulos & Panopoulos 2022).

The German project "Industry 4.0" (IR4.0) has gained worldwide traction in the last 10 years. Similar strategic projects have been launched by other nations, and a significant amount of research has gone into creating and deploying some of the Industry 4.0 technologies. Industry 5.0

(IR5.0) is driven by values (Rosak-Szyrocka, Żywiołek & Shahbaz 2023), whereas Industry 4.0 is said to be driven by technology (Rosak-Szyrocka & Tiwari 2023; Singh, Rosak-Szyrocka & Tamàndl 2023; Xu et al. 2021). The Operator 4.0 generation embodies the "operator of Industry 4.0," a knowledgeable and proficient operator who, when necessary, uses machines to assist in "work." By integrating human–cyber–physical systems, it provides a new design and engineering philosophy for adaptive production systems that treats automation as a way to improve human physical, sensorial, and cognitive capacities (Mourtzis, Angelopoulos & Panopoulos 2022).

With manufacturing that respects the limits of our planet and puts the welfare of industry workers at the core of the production process, Industry 5.0 acknowledges the capacity of industry to fulfill social objectives beyond employment and development, to become a resilient supplier of wealth. The shift to a sustainable, human-centered, and resilient European industry is driven by research and innovation under Industry 5.0, which is a complement to the current Industry 4.0 paradigm (Industry 5.0: towards a sustainable, human-centric and resilient European industry 2021; Nahavandi 2019). Industry 5.0—also known as the "Age of Augmentation"—is characterized by the harmonious coexistence of humans and machines (Longo, Padovano & Umbrello 2020; Özdemir & Hekim 2018). To overcome challenges and develop novel, economical solutions for guaranteeing the long-term sustainability of manufacturing operations and the welfare of the workforce in the face of challenging and/or unforeseen circumstances, Operator 5f *Industry 5.0: towards a sustainable, human-centric and resilient European industry* 2021.0 is characterized as a knowledgeable and proficient operator who leverages human creativity, ingenuity, and innovation, supported by information and technology (Mourtzis, Angelopoulos & Panopoulos 2022). Three interrelated basic values—human-centricity, sustainability, and resilience—are at the heart of Industry 5.0 (Figure 1.2).

A complete *human-centric* and society-centric strategy replaces technology-driven development by placing fundamental human wants and interests at the center of the manufacturing process. Because of this, people in the sector will take on new tasks as the value of humans shifts from "cost" to "investment." As technology exists to serve people and society, it is adaptable to the various demands and backgrounds of those employed in the industrial sector. To protect workers' basic rights—autonomy, human dignity, and privacy—a safe and inclusive work environment must be established that prioritizes physical, mental, and overall welfare. For greater career options and a work–life balance, industrial workers must continually retrain and upskill themselves.

The industry has to be *sustainable* to respect limits on the planet. It must establish circular processes that decrease waste and their negative effects on the environment, reuse, repurpose, and recycle natural resources to create a circular economy that uses resources more effectively and efficiently.

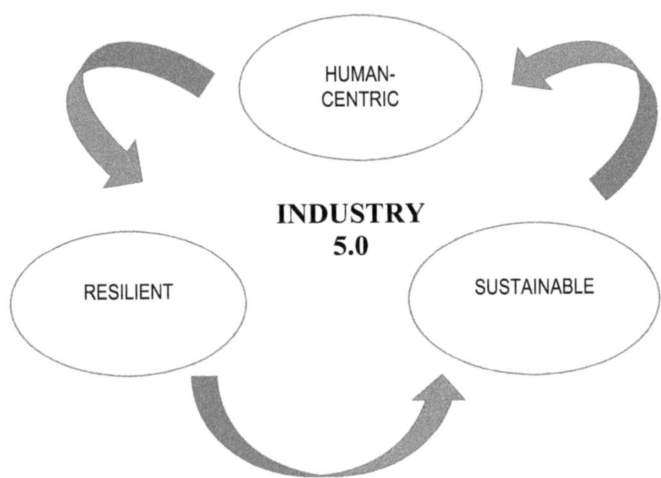

Figure 1.2 Core values of Industry 5.0.

Source: Own study based on: Industry 5.0: towards a sustainable, human-centric and resilient European Industry (2021).

The necessity for industrial production to become more *resilient* is referred to as resilience. This will better equip it to withstand interruptions and guarantee that it can supply and sustain vital infrastructure during emergencies. Natural disasters and (geo-)political upheavals must be quickly navigated by the future industry.

The next "age," the Age of Augmentation (Industry 5.0), will be focused on the collaboration between human intelligence and cognitive computing and on treating automation as a further enhancement of the human physical, sensorial, and cognitive capabilities. The previous two "industrial ages," the Age of Automation (Industry 3.0) and the Digital Age (Industry 4.0), saw automated manufacturing processes and workflows, because of increasingly sophisticated connected devices and computer technologies (Romero et al. 2016).

The Fifth Industrial Revolution, which brought human laborers back to the factory floors, combined workflows with intelligent technology to further use human ingenuity and brainpower to improve process efficiency. Industry 5.0 is the result of the cooperation of autonomous machines and people. The workforce that operates autonomously will possess a keen sense of human purpose and desire. Humans will coexist with robots in the workplace, not just fearlessly but also with confidence, knowing that their robotic colleagues will be able to communicate and work together with humans in an efficient manner. It will lead to a highly productive and value-added manufacturing process, thriving trusted autonomy, and a decrease in waste and related expenses. Industry 5.0 will modify what constitutes a "robot." Not only will robots be programmed machines capable of carrying out monotonous jobs, but in some situations, they will also be perfect human partners. The next industrial revolution will bring in the next generation of robots, known as cobots, that will already know what to perform or rapidly pick it up, allowing robotic products with a human touch. As these cooperative robots are aware of human presence, they will attend to safety and risk requirements. They are able to see, comprehend, and sense not only the human person but also the objectives and standards of a human operator. Similar to an apprentice, a cobot will observe someone else to pick up on their methods. The cobots will carry out the intended duties much like their human operators after they have gained knowledge. Thus, while working with cobots, people sense a new kind of satisfaction (Nahavandi 2019). The literature review also identified five trends that Industry 5.0 is headed toward, including:

- Industry 5.0 in the context of business management, innovation, and digitalization;
- Industry 5.0 in the context of intelligent and sustainable manufacturing;
- Industry 5.0 in the context of evaluating and streamlining the supply chain for industrial operations;
- Industry 5.0 transformation powered by Internet of Things (IoT), Big data, and artificial intelligence (AI); and
- Human–machine connectivity and coexistence (Akundi et al. 2022).

There are now two ideas about Industry 5.0. "Human–robot co-working" comes first. Humans and robots will collaborate whenever and wherever it is feasible under this scenario. Robots will handle the repetitive duties while humans concentrate on those needing ingenuity. The bioeconomy is another idea for Industry 5.0 (Demir & Cicibaş 2018; Ordieres-Meré, Gutierrez & Villalba-Díez 2023). An equilibrium between the economy, industry, and environment may be achieved via the wise utilization of biological resources for industrial uses. "The production of renewable biological resources and the conversion of these resources and waste streams into value-added products, such as food, feed, bio-based products, and bioenergy" is what the European Commission defines as the bioeconomy. It covers portions of the chemical, biotechnology, and energy sectors in addition to agriculture, forestry, fishing, food, and pulp and paper manufacturing. "The utilization of a broad spectrum of sciences (life sciences, agronomy, ecology, food science, and social sciences), industrial and enabling technologies (biotechnology, nanotechnology, information and communication technologies (ICT), and engineering), as well as local and tacit knowledge, gives its sectors a strong potential for innovation" (2012; Sindhwani et al. 2022). The bioeconomy's guiding concept, biologization, has the power to drastically alter several sectors (Schütte 2018). Thus, the bioeconomy might be the focus of the next industrial revolution, or at least a component of it.

1.2.2 Customer 5.0 and Society 5.0 in the era of personalization and digitalization

In the past, customers were only spectators of the material that producers offered, but now they also collaborate with producers to create their own content. Producers are driven by this shift to seek out novel business strategies that may provide extra value for clients (Carvalho & Alves 2023; Rosak-Szyrocka & Żywiołek 2022). Industry 5.0 offers mass customization as part of a new marketing paradigm; in the process, a brand-new class of "super-empowered customer," or "consumer 5.0," has emerged. Society 5.0 is now a part of the new Industry 5.0 era. Some have referred to civilization 5.0 as a "super-intelligent society" or a "humane society" (Rosak-Szyrocka & Tiwari 2023). According to some, Industry 5.0 is a dynamic process of technical advancement. The advent of Industry 5.0 has brought about a paradigm shift in corporate practices from traditional profit-driven and consumption-driven models to more resilient, circular, sustainable, and regenerative models that create value (Sarıoğlu 2023). To hasten the adoption of digital technology, Society 5.0 aims for everyone to actively participate in the integration of these technologies into a variety of procedures (Da Costa Tavares & do Carmo Azevedo 2021). In contrast to many previous ideas, society now approaches and regulates social issues differently. According to Da Costa Tavares and do Carmo Azevedo (2021) and Nair, Tyagi and Sreenath (2021), society has advanced from many of these previous notions (information) with such creative endeavors as Society 1.0 (hunger gatherer), Society 2.0 (agricultural), Society 3.0 (industrialized), and Society 4.0. It is expected that the systems of the social and virtual worlds would be able to be fixed and any holes filled throughout the Society 5.0 timeframe. The development of science and technology that mainly meets the demands of the economics sector is given top priority by Society 5.0 to further humanity (Nugraha & Aminur Rahman 2021). The attributes of Society 5.0 are as follows: (1) full use of information and communication technology; (2) community emphasis; (3) human interaction; (4) common goals of sustainability, inclusiveness, efficacy, and intellectual power; and (5) expansion of commercial disruption. Maximizing the potential of individual technological interactions to achieve societal benefit is emphasized in Sustainability in Society 5.0 (Nugraha & Aminur Rahman 2021). Digital change contributes to the development of a new society with distinct societal traits and ways of living. According to Fujii, Guo and Kamoshida (2018), Society 5.0, or the super smart society, heralds a new era for the social structure. In Society 5.0, creativity is valued highly and is necessary to recognize and address people's wants and issues. Because of its creative and imaginative digital development, Society 5.0 is sometimes referred to as a "Creative Society" (Carvalho & Alves 2023; Demir, Döven & Sezen 2019).

The need for small-batch, customized goods is becoming more and more pressing for the industrial sector. Stated differently, manufacturers must fulfill a variety of client demands via individualization and realize scale effects throughout the value chain (Yavari & Pilevari 2020; Katoozian & Zanjani 2022; Machado et al. 2023). In the twenty-first century, we are dealing with a prosumer who has extensive knowledge about products and services related to a given brand, which they pass on to others. Prosumers are also more aware of their role in relations with producers. This awareness has evolved from formal confirmation of systems to actual market and organizational expectations (Szyrocka et al. 2023). Consumers who are prosumers are more engaged, picky, and conscientious (Mazurek & Tkaczyk 2016). Consumer engagement and consumption are linked. Prosumers actively develop new goods and services, improve materials, increase the dependability and longevity of products, and design packaging and other manufacturing processes (Bartosik-Purgat & Filimon 2023). Customers find it appealing to have the option to participate in the manufacturing process as it will engage them in the company's operations. The younger generations find this tendency more appealing. Using social media and modern IT might make this notion possible (Bednarz 2023).

Prosumers participate more actively in the creation and development of products than do digital and e-consumers. Customers want mass customization and personalization; thus, no robot or other machine in the sector can take the position of human elements (Lu et al. 2021). E-consumers live in virtual worlds and make purchases online. However, as they have access to digital material, they are not limited to making purchases online. Digital consumers may be classified as either passive (by viewing websites) or active (by blogging or leaving comments) (Mazurek & Tkaczyk 2016). Digital customers now concentrate more on value creation as a consequence of the digital revolution, and new consumers are now co-creators of value. Consumers are now more actively involved in value creation, and firms and consumers are working together more often (George & Paul 2020; İpek 2020). The revised prosumer value propositions are shown in Figure 1.3.

A new kind of "super-empowered customer" is the outcome of Industry 5.0's proposed mass-customization marketing paradigm. Maddikunta et al. (2022) state that hyper-customization inside the manufacturing processes will satisfy each customer's customized production solutions using AI-based cognitive systems. Customers may design and produce one-of-a-kind, highly customized products and services using digital devices. New customers will benefit from realistic and captivating encounters as well as personalized

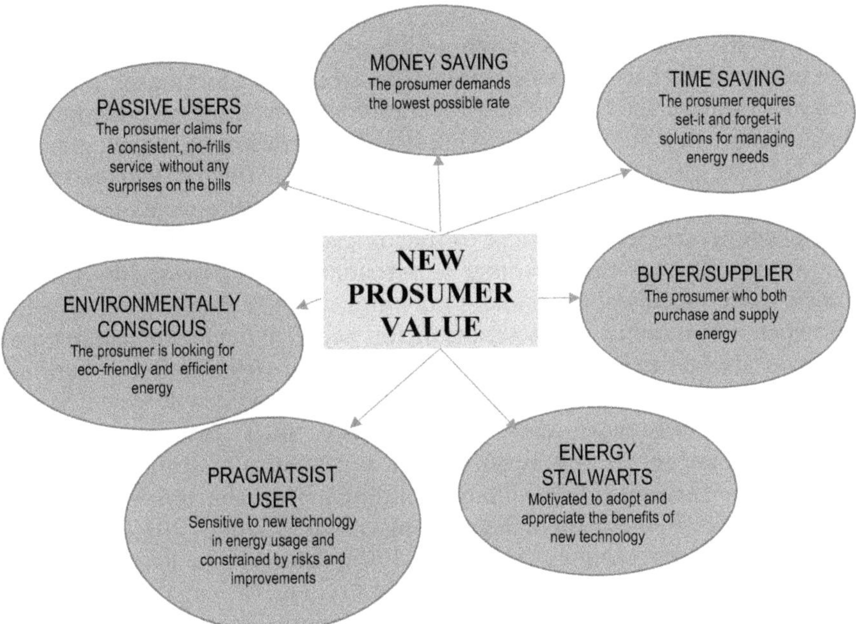

Figure 1.3 New prosumer value.

Source: Own study based on: Sarıoğlu (2023).

customer experiences, because of the marriage of creativity and high-tech (Adel 2022). With the use of digital devices, service providers may leverage changes in consumer behavior to enhance real-world marketing effectiveness by using the virtual world, which contains all of the customer's experiences (Leng et al. 2022).

Nowadays, customers expect continuous improvement of products and services in an increasingly competitive market. From the point of view of enterprises, not only the economic result becomes important, but also the demonstration of the organization's social role and the way it treats both consumers (external customers) and employees (internal customers). Their awareness means that companies must listen to the preferences and tastes of customers and their changing lifestyles. A satisfied customer is a loyal customer, a "living" advertisement for the organization; it improves the company's image and also improves the company's efficiency. Without their satisfaction, their loyalty and fidelity to the company cannot be expected. Enterprises know that they must make every effort to place the customer at the "heart" of the business model to obtain customer satisfaction, which is reflected in the offer profile. The future of competition is entirely related to a new approach to value creation based on the co-creation of value by the company's customers. The essence of the customer as a co-creator of the company's value is associated with moving away from separating the function and management of production from the function of consumption and treating the market as a forum for the exchange of value between the company and the customer. This close cooperation between customers and the enterprise must be based on the commitment of both parties to build value, and the value of the enterprise will increase as co-creation experiences increase. According to the authors (*Ramaswamy v* 2005), the quality of value for the customer is determined and assessed by the customer, while successes in the quality of value delivery depend on individual customer assessments. The condition for acquiring and retaining a customer is a system that determines the quality of the business. There is a need to look for the so-called dynamic customers who will force the supplier to constantly learn and care for quality. Such customers are called "lighthouse keepers," and meeting their needs helps manufacturers move towards the future and find better solutions (Can the hospitality sector ask customers to help them become more sustainable? 2017). Research conducted by the authors (Skrzypek 2000; Ulewicz & Rosak-Szyrocka 2017) shows that quality is a co-responsibility, i.e., the organization's maturation in terms of quality, i.e., the degree of maturity of undertaken activities. Striving for organizational maturity, especially qualitative maturity, helps in development and stimulates the development and success of the enterprise. It allows for achieving measurable economic benefits as well as those that support the creation of value reflected in the offered quality that is appreciated by the customer (Skrzypek 2000). Currently, customers can compare the offers of competing entities located in distant places, because of the various distribution channels of their products provided by the offerer, and thus make decisions more freely. Customers are no longer perceived as passive recipients of the marketing activities of market entities but become their active partners and "advocates." They are also often aware of the impact of their decisions on the specific business goals of companies. Without the acceptance of enterprises' actions by customers, for example, in the area of choosing a distribution channel, enterprises may not achieve the expected revenues and thus exist and develop on the market. The value that the customer receives with the offered product includes four categories of benefits: benefits related to the features and properties of a given product related to its functioning, price, time; benefits related to its operation; benefits resulting from additional amenities; and benefits resulting from prestige and trust. The customer determines the value received as a result of using a given product by comparing the total value (the benefits he received from using the product) with the total cost he incurred. The value delivered to the customer can be calculated using the following formula (Carvalho & Alves 2023):

$$\text{Customer value} = \text{total customer value} - \text{total cost to the customer} \tag{1.1}$$

Total value for the customer is the sum of all values related primarily to the product but also to service during sales and throughout the entire period of use of the product, quality of service, warranty, and other accompanying services; value is also perceived through the prism of factors such as the quality of contacts with staff and the brand or image of the company. The total cost is the sum of all costs related to the purchase of a specific offer, primarily monetary but also other, much more difficult to determine, such as time and energy, that must be incurred in connection with the purchase and use of the product. The lower the costs in relation to the benefits, the higher the value from the customer's point of view, and the better the quality of the purchased product will be assessed.

Automation and AI processes are used in Industry 5.0 to generate the precise goods and services that consumers want. Because of increased flexibility and the ability to fulfill new consumers because of innovative manufacturing, clients may define exactly what they want to purchase (Aslam et al. 2020). According to Grewal et al. (2020), personalization in marketing allows one to promptly and affordably address the unique demands of every consumer.

In terms of marketing, Industry 5.0 entails a transition from a phase of mass customization to one of mass personalization with the use of intelligent data consumption data for sustainable development (Tiwari, Bahuguna & Walker 2022). When it comes to product and service development, mass customization takes into account the unique wants of each consumer. Industry 5.0 provides an atmosphere that is conducive to ensuring the satisfaction of each and every consumer. Personalized services that cater to clients' requirements and lifestyles have replaced the old notion of a great customer, which was centered on amiable salespeople. One example is how Netflix caters to and supports its users' needs and wants by providing them with content that suits their interests (Lee & Lee 2020).

It is intended for emerging technologies to assist Industry 5.0 in production processes so that clients may get personalized and customized goods and services (Maddikunta et al. 2022). As a result, Industry 5.0 seeks to provide the greatest possible client experience. Throughout the whole process, from the supply chain to the marketing initiatives, advanced digital marketing methods using AI, VR, and machine learning personalization are used. To provide consumers with personalized service, Industry 5.0 offers a wide range of tools across all domains.

1.3 METHODOLOGY AND ANALYSIS

To consider why companies implement Industry 5.0, a traditional quality management tool was used—the Ishikawa diagram (Figure 1.4) (Assessing

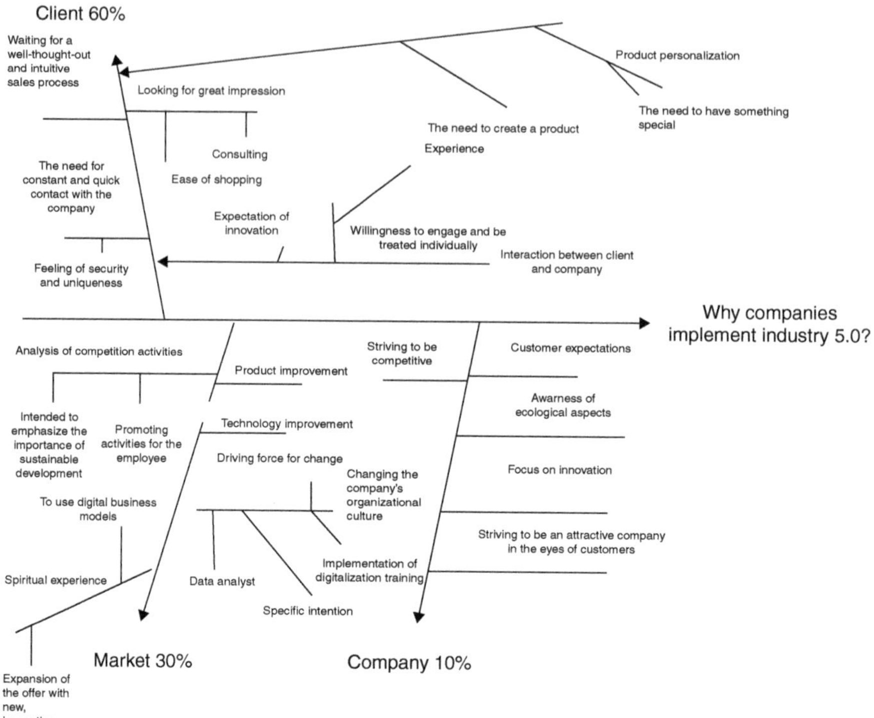

Figure 1.4 Ishikawa diagram base on the question: why companies implement Industry 5.0?

port sustainability in Ecuador using an Ishikawa Diagram 2023; Barsalou & Starzynska 2023; Ghobakhloo et al. 2023; Rosak-Szyrocka & Żywiołek 2022). One of the classic quality management instruments that is often used by different management systems is the Ishikawa diagram. Ishikawa is used to ascertain a problem's cause-and-effect linkages. It is predicated on a graphical depiction of the investigation of the connection between the factors that lead to a certain issue. The instrument aids in identifying the source of the issue (Rosak-Szyrocka & Żywiołek 2022). The question that follows was examined: Why do companies implement Industry 5.0? The primary determinants impacting the execution have been examined. Businesses adopt Industry 5.0 primarily for three reasons: (1) customers; (2) markets; and (3) businesses. One of the main elements in fostering a favorable impression of the business in the eyes of the client—which often leads to a purchase—is personalization. According to a study conducted by Gartner, if the seller did not adjust the offer and message to their satisfaction, as much as 54% of respondents would give up on the transaction.

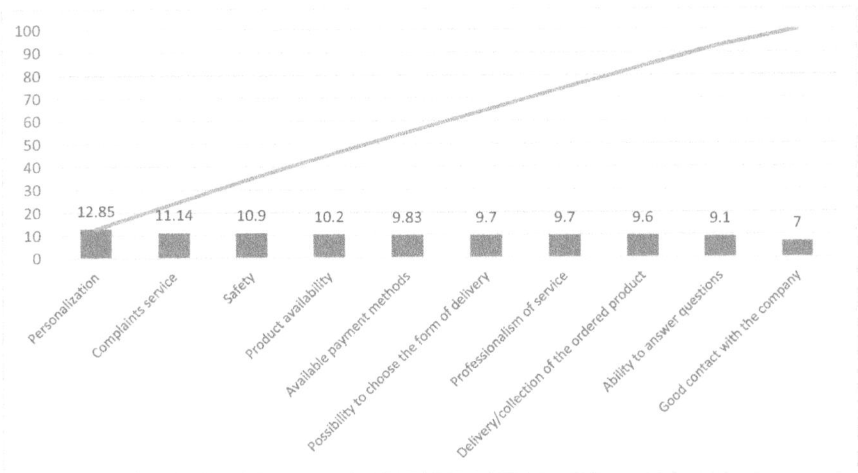

Figure 1.5 Pareto–Lorenz diagram to analyze how customers perceive value.

Thus, a growing number of vendors choose to include customization into their business procedures. A new age of customization for both consumers and staff is made possible by Industry 5.0.

In the following part, the Pareto–Lorenz diagram was used (Figure 1.5) to analyze how customers perceive value (Rosak-Szyrocka & Żywiołek 2022). The research was conducted using the computer-assisted web interviewing (CAWi) survey in 2021 among 404 respondents, of whom 62% were women and the remaining 33% were men. Analysis of the Pareto–Lorenz diagram shows that 20% of the factors consisting of Personalization and Complaints service determines 23.99% of the customer's value in the customer's opinion.

1.4 DISCUSSION AND CONCLUSION

Industry 5.0 is a values-driven program that propels technical change toward a specific goal rather than a revolution driven by technology (Industry 5.0: towards a sustainable, human-centric and resilient European industry 2021). The concept of Industry 5.0 emerged from a forward-thinking effort that sought to define the coexistence of industry with newly developing society trends and requirements. Thus, Industry 5.0 expands upon and enhances the defining characteristics of Industry 4.0. Human–machine interaction (HMI) will face hitherto unheard-of difficulties as a result of Industry 5.0, which will bring robots closer to people's daily lives. Industry 5.0 aims to eliminate tedious, unclean, and repetitive jobs from human labor whenever feasible, thereby transforming production processes worldwide (Pizoń & Gola 2023; Zhang et al. 2023). Manufacturing supply networks and factory floors will see an unprecedented infiltration of intelligent robots and systems. The advent of less expensive and more powerful robots built from cutting-edge materials like carbon fiber and lightweight but durable materials, powered by highly optimized battery packs, cyberattack resistant, with more robust data handling procedures (such as Big data and AI), and an intelligent sensor network will make this feasible. Industry 5.0 will be more environmentally friendly, minimize work-related injuries, shorten production cycles, and boost productivity and operational efficiency. Nevertheless, Industry 5.0 will generate more employment than it eliminates, despite first perception to the reverse. There will be a significant increase in employment opportunities in the field of intelligent systems, including programming, maintenance, training, scheduling, repurposing, and the development of new types of industrial robots. Furthermore, by encouraging everyone to creatively employ various types of robots in the workplace, repetitive jobs that would otherwise need human labor may be completed by robots, thereby boosting creativity in the work process (Alves, Lima & Gaspar 2023; Ghobakhloo et al. 2022; Nahavandi 2019). Industry 5.0 is replacing earlier advancements, and because of its extreme precision and the reduction of human labor time and effort via machine work, it is an efficient process. A few qualities, apart from the difficulties, motivate business organizations to adopt Industry 5.0. The majority of industries are heading toward the smart social factory by implementing Industry 5.0 (Mourtzis 2022). In addition to attempting to connect communication across the supply chain, the plant, the end-user, and transportation, Industry 5.0 concentrates on establishing a tight relationship between the process and cobots (coordinated robots). Massively collected, transmitted, analyzed, and stored data sets the stage for low-cost manufacturing breakthroughs and digital transformation in several domains (Mamasioulas, Mourtzis & Chryssolouris 2020). Industry 5.0's main goal is to build a better society where employees are content with their lives and are prepared for increased productivity and automated procedures. The move to a natural environment, user-based customization, integration of the cognitive domain into human intelligence, and penalization are all features of Industry 5.0. The creation of inventive societies is thus the driving force behind Industry 5.0, and the interaction between humans and robots using repeatable energy is the key technology of this movement (Verma et al. 2022).

A key component of the Industry 5.0 process' digital user experience is mass customization. A key production paradigm known as "mass customization" suggests creating customized goods and services in large quantities with prices that are comparable to those of mass manufacturing (Bednarz 2023). Reconfigurable production systems and product family architecture are examples of emerging technologies that provide mass customization while delaying differentiation (Hu 2013). The primary focus of Industry 5.0

is the reintroduction of human components into the manufacturing scene, namely human contact. AI plays a pivotal role in this paradigm (Grabowska, Saniuk & Gajdzik 2022).

Hyper-personalization is the process of tailoring product and service experiences to individual user preferences by collecting real-time data on customer behavior. To establish a shared knowledge of the particular products and services, the characteristics of the consumers, and the technology behind the process, the industrial production area also places hyper-personalization at the center of the system. Personalized marketing methods ensured industrial processes for particular items by using computer vision, machine learning, and AI to gain knowledge of digital technologies (Carvalho & Alves 2023). Industry 5.0 will increase and improve customers' awareness of social and environmental concerns when they make decisions about what to buy (Nahavandi 2019).

REFERENCES

2012, 'Innovating for sustainable growth: a bioeconomy for Europe', *Industrial Biotechnology*, vol. 8, no. 2, pp. 57–61. Available from: https://op.europa.eu/en/publication-detail/-/publication/1f0d8515-8dc0-4435-ba53-9570e47dbd51

2017, 'Can the hospitality sector ask customers to help them become more sustainable?' Available from: www.researchgate.net/publication/310797575_Can_the_hospitality_sector_ask_customers_to_help_them_become_more_sustainable

2021, 'Industry 5.0: towards a sustainable, human-centric and resilient European industry'. https://research-and-innovation.ec.europa.eu/news/all-research-and-innovation-news/industry-50-towards-more-sustainable-resilient-and-human-centric-industry-2021-01-07_en

2023, 'Assessing port sustainability in Ecuador using an Ishikawa Diagram'. Available from: https://worldscientificnews.com/assessing-port-sustainability-in-ecuador-using-an-ishikawa-diagram/

Adel, A 2022, 'Future of Industry 5.0 in society: human-centric solutions, challenges and prospective research areas', *Journal of Cloud Computing*, vol. 11, no. 1, p. 40. Available from: https://journalofcloudcomputing.springeropen.com/articles/10.1186/s13677-022-00314-5.

Akundi, A, Euresti, D, Luna, S, Ankobiah, W, Lopes, A & Edinbarough, I 2022, 'State of Industry 5.0—analysis and identification of current research trends', *Applied System Innovation*, vol. 5, no. 1, p. 27. Available from: www.mdpi.com/2571-5577/5/1/27.

Alves, J, Lima, TM & Gaspar, PD 2023, 'Is Industry 5.0 a human-centred approach? A systematic review', *Processes*, vol. 11, no. 1, p. 193. Available from: www.mdpi.com/2227-9717/11/1/193.

Aslam, F, Aimin, W, Li, M & Ur Rehman, K 2020, 'Innovation in the era of IoT and Industry 5.0: absolute innovation management (AIM) framework', *Information*, vol. 11, no. 2, p. 124. Available from: www.mdpi.com/2078-2489/11/2/124.

Barsalou, M & Starzynska, B 2023, 'A new method for formulating a strong hypothesis in RCA', *International Journal for Quality Research*, vol. 17, no. 1, pp. 77–96.

Bartosik-Purgat, M & Filimon, N 2023, *European Consumers in the Digital Era. Implications of Technology, Media and Culture on Consumer Behavior*, Routledge, New York.

Bednarz, J 2023, 'Who is a consumer in the digital era? Still a consumer or a pro-sumer?' in *Digital Consumer Behaviour in Europe. Implications of Technology, Media and Culture on Consumer Behavior*, eds M Bartosik-Purgat & N Filimon, Routledge, New York, pp. 27–42.

Carvalho, P & Alves, H 2023, 'Customer value co-creation in the hospitality and tourism industry: a systematic literature review', *International Journal of Contemporary Hospitality Management*, vol. 35, no. 1, pp. 250–273.

Da Costa Tavares, MC & do Carmo Azevedo, GM 2021, 'Society 5.0 as a contribution to the sustainable development report' in *Advances in Tourism, Technology and Systems. Selected papers from ICOTTS20*, eds JV de Carvalho, A Rocha, P Liberato & A Peña, Springer Singapore, Singapore, pp. 49–63. Available from: https://link.springer.com/chapter/10.1007/978-981-33-4256-9_5.

Demir, KA & Cicibaş, H 2018, 'The next industrial revolution: Industry 5.0 and discussions on Industry 4.0' in *Industry 4.0 from the Management Information Systems Perspectives*. Peter Lang GmbH, Internationaler Verlag der Wissenschaften. Available from: www.researchgate.net/publication/336653504_The_Next_Industrial_Revolution_Industry_50_and_Discussions_on_Industry_40

Demir, KA, Döven, G & Sezen, B 2019, 'Industry 5.0 and Human–Robot Co-Working', *Procedia Computer Science*, vol. 158, pp. 688–695. Available from: www.sciencedirect.com/science/article/pii/s1877050919312748.

Fujii, T, Guo, T & Kamoshida, A 2018, 'A consideration of service strategy of Japanese electric manufacturers to realize super smart society (SOCIETY 5.0)' in *Knowledge Management in Organizations. Proceedings*, eds L Uden, B Hadzima & I-H Ting, Springer, Cham, pp. 634–645. Available from: https://link.springer.com/chapter/10.1007/978-3-319-95204-8_53.

George, BP & Paul, J (eds.) 2020, *Digital Transformation in Business and Society. Theory and Cases*, Palgrave Macmillan, Cham, Switzerland.

Ghobakhloo, M, Iranmanesh, M, Mubarak, MF, Mubarik, M, Rejeb, A & Nilashi, M 2022, 'Identifying Industry 5.0 contributions to sustainable development: a strategy roadmap for delivering sustainability values', Sustainable Production and Consumption, vol. 33, pp. 716–737. Available from: www.sciencedirect.com/science/article/pii/s2352550922002093.

Ghobakhloo, M, Iranmanesh, M, Tseng, M-L, Grybauskas, A, Stefanini, A & Amran, A 2023, 'Behind the definition of Industry 5.0: a systematic review of technologies, principles, components, and values', *Journal of Industrial and Production Engineering*, vol. 40, no. (6), pp. 432–447.

Grabowska, S, Saniuk, S & Gajdzik, B 2022, 'Industry 5.0: improving humanization and sustainability of Industry 4.0', *Scientometrics*, vol. 127, no. 6, pp. 3117–3144. Available from: https://link.springer.com/article/10.1007/s11192-022-04370-1.

Grewal, D, Hulland, J, Kopalle, PK & Karahanna, E 2020, 'The future of technology and marketing: a multidisciplinary perspective', *Journal of the Academy of Marketing Science*, vol. 48, no. 1, pp. 1–8. Available from: https://link.springer.com/article/10.1007/s11747-019-00711-4.

Higher Education and Digitalization in Perspective of Use of Internet, Integration of Digital Technology, Digital Public … 2023. Available from: www.researchgate.net/profile/joanna-rosak-szyrocka/publication/376232419_higher_education_and_digitalization_in_perspective_of_use_of_internet_integration_of_digital_technology_digital_public_services_panel_study_of_eu_nations/links/6571d56fea5f7f02054cc6d8/higher-education-and-digitalization-in-perspective-of-use-of-internet-integration-of-digital-technology-digital-public-services-panel-study-of-eu-nations.pdf.

Hu, SJ 2013, 'Evolving paradigms of manufacturing: from mass production to mass customization and personalization', Procedia CIRP, vol. 7, pp. 3–8. Available from: www.sciencedirect.com/science/article/pii/s2212827113002096.

İpek, İ 2020, 'Understanding consumer behavior in technology-mediated spaces' in *Digital Transformation in Business and Society. Theory and Cases*, eds BP George & J Paul, Palgrave Macmillan, Cham, Switzerland, pp. 169–189. Available from: https://link.springer.com/chapter/10.1007/978-3-030-08277-2_11.

Katoozian, H & Zanjani, MK 2022, 'Supply network design for mass personalization in Industry 4.0 era', *International Journal of Production Economics*, vol. 244, p. 108349. Available from: www.sciencedirect.com/science/article/pii/s092552732100325x?casa_token=bhvs-nb1hauaaaaa:oiqqkm6tgfpktn5fga7g3jop8_ohderrfl4kits78m6tvbkjsh19bcv0f8myxrtdb95vnqhn-ki.

Lee, SM & Lee, D 2020, '"Untact": a new customer service strategy in the digital age', *Service Business*, vol. 14, no. 1, pp. 1–22. Available from: https://bit.ly/4dZEKVd.

Leng, J, Sha, W, Wang, B, Zheng, P, Zhuang, C, Liu, Q, Wuest, T, Mourtzis, D & Wang, L 2022, 'Industry 5.0: prospect and retrospect', *Journal of Manufacturing Systems*, vol. 65, pp. 279–295. Available from: www.sciencedirect.com/science/article/pii/s0278612522001662?casa_token=7x75cqiraycaaaaa:sjuq5h1lo7kzv43t3nijgrbgxamlt7jus1ln3hflmnf6yqxi2imxjspw0iyqgueuasksz_lypclz.

Longo, F, Padovano, A & Umbrello, S 2020, 'Value-oriented and ethical technology engineering in Industry 5.0: a human-centric perspective for the design of the factory of the future', *Applied Sciences*, vol. 10, no. 12, p. 4182. Available from: www.mdpi.com/2076-3417/10/12/4182.

Lu, Y, Adrados, JS, Chand, SS & Wang, L 2021, 'Humans are not machines—anthropocentric human–machine symbiosis for ultra-flexible smart manufacturing', Engineering, vol. 7, no. 6, pp. 734–737. Available from: www.sciencedirect.com/science/article/pii/s2095809921001612.

Machado, J, Soares, F, Trojanowska, J, Ivanov, V, Antosz, K, Ren, Y, Manupati, VK & Pereira, A (eds.) 2023, *Innovations in Industrial Engineering II*, Springer, Portugal.

Maddikunta, PKR, Pham, Q-V, Prabadevi, B, Deepa, N, Dev, K, Gadekallu, TR, Ruby, R & Liyanage, M 2022, 'Industry 5.0: a survey on enabling technologies and potential applications', *Journal of Industrial Information Integration*, vol. 26, p. 100257. Available from: www.sciencedirect.com/science/article/pii/s2452414x21000558?casa_token=vgaii2iv3lcaaaaa:f6xqz2qhttyekhvykmp8e2etzojotjt30j2spntwe6yzrg1lfxr1xuvfcb_atagpusybvljzofhy.

Mamasioulas, A, Mourtzis, D & Chryssolouris, G 2020, 'A manufacturing innovation overview: concepts, models and metrics', *International Journal of Computer Integrated Manufacturing*, vol. 33, no. 8, pp. 769–791.

Mazurek, G & Tkaczyk, J 2016, *The Impact of the Digital World on Management and Marketing*, Poltext, Kozminski University, Warsaw.

Mitrofanova, E, Mitrofanova, A & Margarov, G 2023, 'HRM-ecosystem of the organization in the context of global challenges' in *Proceedings of Computer Science and Information Technologies 2023 Conference*, Institute for Informatics and Automation Problems.

Mourtzis, D 2022, 'The mass personalization of global networks' in *Design and Operation of Production Networks for Mass Personalization in the Era of Cloud Technology*, ed. D Mourtzis, Elsevier, pp. 79–116. https://doi.org/10.1016/B978-0-12-823657-4.00006-3. Available from: www.sciencedirect.com/science/article/pii/b9780128236574000063.

Mourtzis, D, Angelopoulos, J & Panopoulos, N 2022, 'Operator 5.0: a survey on enabling technologies and a framework for digital manufacturing based on extended reality', *Journal of Machine Engineering*, vol. 22, no. 1, pp. 43–69. Available from: https://yadda.icm.edu.pl/baztech/element/bwmeta1.element.baztech-159441e0-80bf-4fed-8670-f0d61c3439f1.

Nahavandi, S 2019, 'Industry 5.0—a human-centric solution', *Sustainability*, vol. 11, no. 16, p. 4371. doi:10.3390/su11164371

Nair, MM, Tyagi, AK & Sreenath, N 2021, 'The future with Industry 4.0 at the core of Society 5.0: open issues, future opportunities and challenges' in 2021 International Conference on Computer Communication and Informatics (ICCCI), IEEE, India.

Nugraha, A & Aminur Rahman, F 2021, 'Android application development of student learning skills in era Society 5.0', *Journal of Physics: Conference Series*, vol. 1779, no. 1, p. 12014. Available from: https://iopscience.iop.org/article/10.1088/1742-6596/1779/1/012014/meta.

Ordieres-Meré, J, Gutierrez, M & Villalba-Díez, J 2023, 'Toward the Industry 5.0 paradigm: increasing value creation through the robust integration of humans and machines', *Computers in Industry*, vol. 150, p. 103947. Available from: www.sciencedirect.com/science/article/pii/s0166361523000970.

Özdemir, V & Hekim, N 2018, 'Birth of Industry 5.0: making sense of Big Data with artificial intelligence, "The Internet of Things" and next-generation technology policy', *OMICS: A Journal of Integrative Biology*, vol. 22, no. 1, pp. 65–76.

Pan, M, Zhao, X, lv, K, Rosak-Szyrocka, J, Mentel, G & Truskolaski, T 2023, 'Internet development and carbon emission-reduction in the era of digitalization: where will resource-based cities go?', *Resources Policy*, vol. 81, p. 103345. doi:10.1016/j.resourpol.2023.103345

Pizoń, J & Gola, A 2023, 'Human–machine relationship—perspective and future roadmap for Industry 5.0 solutions', *Machines*, vol. 11, no. 2, p. 203. Available from: www.mdpi.com/2075-1702/11/2/203.

Romero, D, Bernus, P, Noran, O, Stahre, J & Fast-Berglund, Å 2016, 'The Operator 4.0: human cyber-physical systems & adaptive automation towards human-automation symbiosis work systems' in Advances in Production Management Systems. Initiatives for a Sustainable World. IFIP WG 5.7 International Conference, APMS 2016, Iguassu Falls, Brazil, September 3–7, 2016, Revised Selected Papers, eds I Nääs, O Vendrametto, J Mendes, R Goncalves, M Terra, G von Cieminski & D Kiritsis, Springer International Publishing, Cham, pp. 677–686. Available from: https://link.springer.com/chapter/10.1007/978-3-319-51133-7_80.

Rosak-Szyrocka, J & Tiwari, S 2023, 'Structural Equation Modeling (SEM) to Test Sustainable Development in University 4.0 in the Ultra-Smart Society Era', *Sustainability*, vol. 15, no. 23, p. 16167.

Rosak-Szyrocka, J & Żywiołek, J 2022, 'Qualitative analysis of household energy awareness in Poland', *Energies*, vol. 15, no. 6, p. 2279. Available from: www.mdpi.com/1996-1073/15/6/2279.

Rosak-Szyrocka, J, Apostu, S & Akkaya, B 2023, 'Higher education and digitalization in perspective of use of Internet, integration of digital technology, digital public services: panel study of EU nations', *Scientific Papers of Silesian University of Technology Organization and Management Series*, vol. 2023, pp. 469–491. Available from: https://bit.ly/3VcwXvX.

Rosak-Szyrocka, J, Żywiołek, J & Shahbaz, M 2023, *Quality Management, Value Creation and the Digital Economy*, Routledge, New York.

Sarıoğlu, Cİ 2023, 'Industry 5.0, digital society, and consumer 5.0' in *Perspectives on Society and Technology Addiction*, eds D Khosrow-Pour, R Sine Nazlı & G Sari, IGI Global, pp. 11–33. www.irma-international.org/chapter/industry-50-digital-society-and-consumer-50/325179/

Schütte, G 2018, 'What kind of innovation policy does the bioeconomy need?', *New Biotechnology*, vol. 40, Pt A, pp. 82–86. Available from: www.sciencedirect.com/science/article/pii/s1871678416326127?casa_token=kx1nz31dh04aaaaa:sz06nestx2yppkuoqviybnzhpzkdyv6ov0sacnzpffvwegjb9k0nvjumtfaejorjq1q7z_mdybpu.

Sindhwani, R, Afridi, S, Kumar, A, Banaitis, A, Luthra, S & Singh, PL 2022, 'Can Industry 5.0 revolutionize the wave of resilience and social value creation? A multi-criteria framework to analyze enablers', *Technology in Society*, vol. 68, p. 101887. Available from: www.sciencedirect.com/science/article/pii/s0160791x22000288.

Singh, S, Rosak-Szyrocka, J & Tamàndl, L 2023, 'Development, service-oriented architecture, and security of blockchain technology for Industry 4.0 IoT application', *HighTech and Innovation Journal*, vol. 4, no. 1, pp. 134–156. Available from: www.hightechjournal.org/index.php/hij/article/view/410.

Skrzypek, E 2000, *Jakość i efektywność*, Wydawnictwo UMCS, Lublin.

Szyrocka, JR, Żywiołek, J, Nayyar, A & Naved, M 2023, *Advances in Distance Learning in Times of Pandemic*, Chapman & Hall/CRC, Boca Raton.

Tiwari, S, Bahuguna, PC & Walker, J 2022, 'Industry 5.0: a macroperspective approach' in *Handbook of Research on Innovative Management Using AI in Industry 5.0*, IGI Global, pp. 59–73. https://doi.org/10.4018/978-1-7998-8497-2.ch004. Available from: www.igi-global.com/chapter/industry-50/291461.

Ulewicz, R & Rosak-Szyrocka, J 2017, 'Rola Modelu EFQM w Postrzeganiu Jakości w Polskich Przedsiębiorstwach', *Przegląd Organizacji*, no. 5, pp. 10–16. Available from: https://przegladorganizacji.pl/artykul/2017/10.3314 1po.2017.05.02.

Verma, A, Bhattacharya, P, Madhani, N, Trivedi, C, Bhushan, B, Tanwar, S, Sharma, G, Bokoro, PN & Sharma, R 2022, 'Blockchain for Industry 5.0: vision, opportunities, key enablers, and future directions', *IEEE Access*, vol. 10, pp. 69160–69199.

Xu, X, Lu, Y, Vogel-Heuser, B & Wang, L 2021, 'Industry 4.0 and Industry 5.0—inception, conception and perception', *Journal of Manufacturing Systems*, vol. 61, pp. 530–535. Available from: www.sciencedirect.com/science/article/pii/s0278612521002119.

Yavari, F. & Pilevari, N 2020, 'Industry revolutions development from Industry 1.0 to Industry 5.0 in manufacturing', *Journal of Industrial Strategic Management*, vol. 5, no. 2. Available from: https://sanad.iau.ir/journal/mgmt/Article/678 926?jid=678926

Zhang, C, Wang, Z, Zhou, G, Chang, F, Ma, D, Jing, Y, Cheng, W, Ding, K & Zhao, D 2023, 'Towards new-generation human-centric smart manufacturing in Industry 5.0: a systematic review', *Advanced Engineering Informatics*, vol. 57, p. 102121. Available from: www.sciencedirect.com/science/article/pii/s14740 34623002495?casa_token=4abappza6k0aaaaa:ospzpchl541x_azivoqe0qxj08 x9skf241o7boi0fubvx2d4cc0p2xzfjufgtoniav_amqcy0u7r.

Chapter 2

Smart manufacturing

Intelligent systems in Industry 5.0

Neha Mishra and Atul Garg

2.1 INTRODUCTION

The upcoming era of advanced technologies in the industry is known as the 5.0 revolution, which emphasises a human-centric approach to achieve harmonious coexistence between individuals and robots. In the current era, technology is seamlessly merging with human intelligence through advancements like artificial intelligence (AI) and the Internet of Things (IoT). Throughout the course of human history, there has been a continuous drive for innovation, growth, and progress. It was through collaboration with industry that humans were able to transition from primitive living conditions in caves and jungles to thriving in the most advanced cities. The growth of industry relied not only on the necessary conditions, but also on the surrounding context and ongoing advancements that were accessible during that era. The advent of computers has brought about a new era of remarkable transformation worldwide. It is fascinating to observe the evolution of computers, from the early generations to the latest advancements in computer systems with AI. However, the rapid growth of the Internet and other networks over the past two decades has had a profound impact on the world. The Internet has had a significant impact on various fields, including education, government, insurance, the stock market, medical, and others (Chaka, 2022).

The information technology sector among all has experienced significant growth and has become a crucial industry globally and provides the most promising employment opportunities. Just like the advancement of computers from the first generation to the fifth generation, the growth of research and industry is fuelled by emerging technologies (Garg et al.2021a; Amin et al.2014). Over the past few decades, numerous innovations have surfaced, reshaping people's thinking and changing user demands and expectations for products. The growth of this industry has wide-ranging effects, impacting not just the company's stakeholders, but also society, the nation, and individuals (Kothari 2020).

DOI: 10.1201/9781032677040-2

2.2 INDUSTRY TRANSFORMATION 1.0 TO 5.0

The advancement of individuals and communities resulted in increased demands for a variety of items. A growing demand for goods and services requires additional resources, such as equipment and labour. From 1.0 to 5.0, the industry had to adopt novel creations, which enabled cutting-edge technologies to be implemented(Ozdemir and Hekim,2018). As per Vaidya, Ambad, and Bhosle (2018), the industrial revolution brought about a wave of innovation and technological advancements that significantly boosted productivity while reducing the need for extensive time and labour. People have become more dependent on different technologies as they have embraced them(Klimuk et al.,2020). Consequently, these technologies and robots have significantly reduced the amount of work performed by humans, which has led to a decrease in direct employment opportunities.

The researchers Sharma and Singh (2020), Sinha et al. (2023), Pilevari and Yavari (2020), and Tavares, Azevedo, and Marques (2022) have extensively studied the evolution of industry from version 1.0 to version 5.0. The beginning of Industry 1.0 can be traced back to England in 1784, when the steam engine was developed. In the year 1870, the 2.0 revolution was fuelled by advancements in steel, electrical energy, the division of labour, and assembly lines and resulted in the emergence of fresh opportunities for different forms of employment. In 1945, the introduction of Electronic Numerical Integrator and Computer (ENIAC), the first digital computer, marked a significant turning point for industries. Moreover, the year 1969 is widely recognised as the year when Industry 3.0 was established. Computers, networking, electronics and information technology, automated goods, and other technological advancements are all playing a significant role in shaping the future of industries.

Various advancements in transportation, manufacturing, and communication were made, including the development and utilisation of new methods and technologies such as the World Wide Web and the Internet. Industry 4.0 is built upon a solid foundation of cutting-edge technologies such as data analytics, linked programmes, cyber–physical systems, cloud computing, robotics, communication through smartphones, social media, and the IoT. In 2011 Industry 4.0 reached its peak in terms of development. However, Industry 5.0 is human-centric and is based on Industry 4.0 with better utilisation of technologies like AI with human skills (Revolution 1830; Pilevari and Yavari 2020; Kothari 2020; Chander et al. 2022; Duggal et al. 2022; Lilhore et al. 2022b. Industry 4.0 encountered many challenges when it came to small and medium-sized enterprises (SMEs) (Moeuf et al. 2018; Ingaldi and Ulewicz 2020; Ko et al. 2020).

The technical comparison is depicted in Table 2.1 and in Figures 2.1 and 2.2.

Table 2.1 Comparisons between revolution of Industry 1.0 to 5.0

Revolution	Revolution Year	Identity	Description	Major Evolutions
1.0	1784	Mechanisation	Steam, coal, and water-based industrial manufacturing	Steam engines
2.0	1870	Electrification	Electrical energy, labour division, mass production and assembly line	Electrical and assembly line
3.0	1969	Automation	Computers, networking, applications of electronics, information technologies, automated products	Information technology, automation
4.0	2011	Digitisation	Data analytics, connected programmes, cyber physical systems, cloud computing, robotic, augmented research	Big data, IoT, robotics
5.0	Current	Personalisation	Cognitive systems, mass customisation, AI, cobots, sustainability, human centric	AI, sustainability etc.

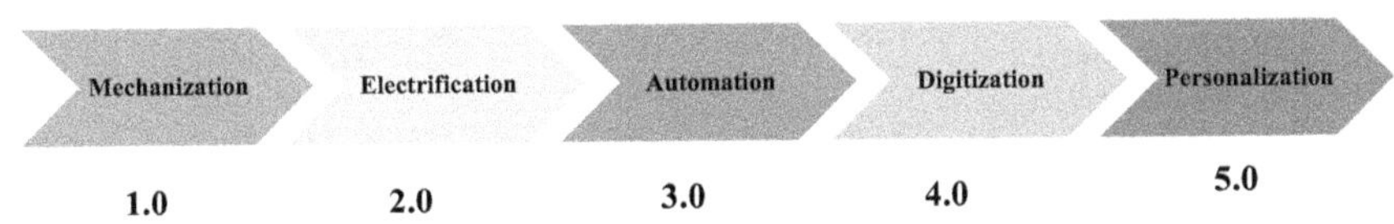

Figure 2.1 Evolution of Industry 1.0 to 5.0.

Source: Own study basis.

2.3 INTEGRATION OF INTELLIGENT SYSTEMS IN SUSTAINABLE INDUSTRIES

AI and robots are already integrated into our daily lives, as highlighted by various studies (Demir, Döven, and Sezen 2019; Martynov, Shavaleeva, and Zaytseva 2019; Ikumapayi et al. 2023) and depicted in Figure 2.3. Doctors at top hospitals in India recommend the use of robot surgery, where humans collaborate with machines to achieve superior outcomes compared to manual labour (Elbeltagi et al. 2020). Robot surgery combines the efforts of individuals and advanced technology to achieve superior results compared to traditional human labour by Elbeltagi et al. (2020) and Aslam et al. (2020). Internet and smartphones have sparked a significant transformation in various industries, such as banking, education, and communication,

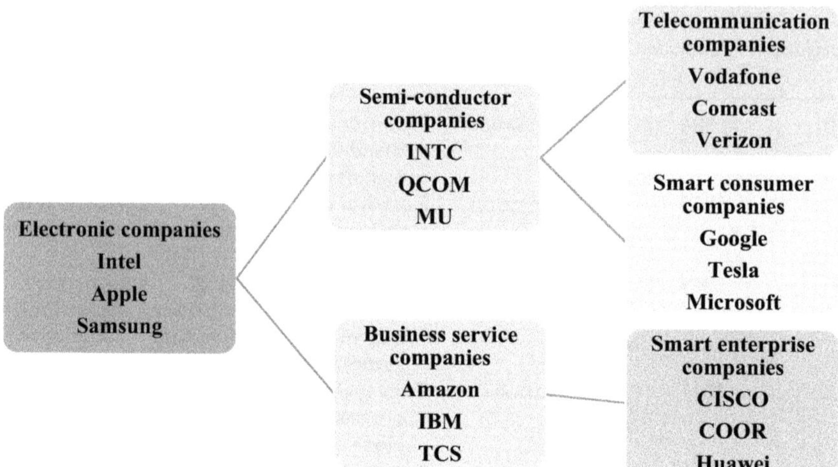

Figure 2.2 Global technology trend.

Source: Own study basis.

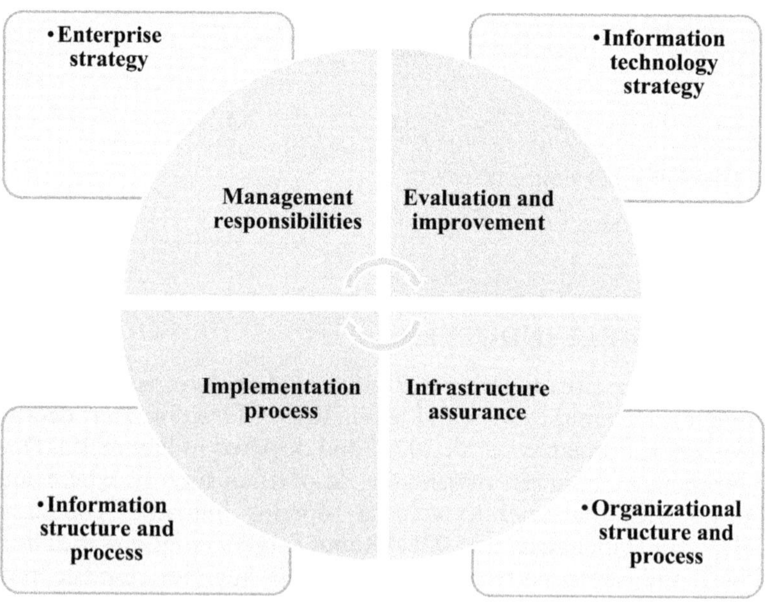

Figure 2.3 Integration of information technology and industrialisation.

Source: Own study basis.

among others. Furthermore, Akundi et al. (2022) have also highlighted examples such as ChatGPT and driverless cars (Emma-Ikata and Doyle-Kent 2022). In addition, sustainability is increasingly becoming a crucial consideration. Throughout the course of corporate history, companies have historically paid little attention to the potential impact of their tactics on future generations. These issues have led to a multitude of problems, including shifts in the climate, contamination of the air and water, economic challenges, and concerns about public health. It is crucial for companies to prioritise functionality over financial gain as one of their top goals. The United Nations (UN) has established 17 Sustainable Development Goals (SDGs) aimed at creating a more prosperous future for generations to come. The primary goal will have wide-ranging effects on the economy, as well as on society and the environment. The connection between Industry 5.0 and environmentally friendly practices has become stronger than ever. Industry 5.0 holds immense promise in its ability to make a substantial positive impact on the environment by conserving natural resources and minimising waste and pollution. As per the study conducted by Gholami et al. (2021), it holds promise in fostering a balanced and fair economy.

2.4 LITERATURE REVIEW

The emerging technology of the preceding period, referred to as "Industry 5.0," is designed to facilitate the operation of efficient and intelligent machinery. Demir, Döven, and Sezen (2019) conducted an analysis indicating that individuals may increasingly depend on robots for the completion of routine domestic chores. However, Demir and Cicibaş (2019) examined the challenges and achievements that have occurred during different industrial revolutions. The research conducted on the intricacies of human–robot interaction has generated significant interest among scholars, prompting them to embark on an examination of robotics as a field of study. Brunetti, Gena, and Vernero (2022) and Ingaldi and Ulewicz (2020) also analysed the challenges faced by SMEs within the framework of Industry 4.0, such as limited availability of things, the high cost of obtaining financial support, and the unfavourable environment. A study was conducted by Ko et al. (2020) in SMEs in Korea and found robots and cobots played significant roles in various industries(Saxena et al., 2021), (Wang et al., 2021), (Huang et al., 2022),(Voulgaridis et al., 2022). Similarly, managing production planning and control can be achieved through the use of an enterprise resource planning (ERP) system. This type of system allows for seamless integration of data in both horizontal and vertical directions. A study conducted by Cimini et al. (2021) indicated that a large number of respondents embraced the latest technologies, such as augmented reality (AR) and virtual reality (VR), for different purposes..

According to Xu et al. (2021), Industry 5.0 distinguishes itself from Industry 4.0 by placing greater emphasis on value creation. The authors also suggested that significant deliberation and careful planning were required prior to ensuring the accessibility of Version 5.0 for users. Wulandari, Winarno, and Triyanto (2021) and Turner et al. (2022) emphasised the significance of maintaining digital ethics and protecting the environment. Jafari, Azarian, and Yu (2022) anticipated advancements in high-performance materials, automation, systems, and devices with embedded intelligence with the implication of cobots and other advanced technology. ElFar et al. (2021) investigated the topics of sustainable electricity, bioenergy, green building, and cleanliness. The progress made in various domains such as technology, algorithms, biofuels, mechanical engineering, and agriculture is facilitating the emergence of Industry 5.0, which aims to address society's sustainability requirements(Wulandri et al.,2021). Several experts, such as Salimova, Vukovic, and Guskova (2020), ElFar et al. (2021), Grabowska, Saniuk, and Gajdzik (2022), and Saniuk, Grabowska, and Straka (2022), have conducted thorough investigations and gathered substantial data on smart implications of Industry 5.0.

Suryadi, Kushardiyanti, and Gusmanti (2021) have explored the extensive impact of Industry 5.0 on the economy. Toward future of industries, the seamless incorporation of digital technology and the immense potential of AI-powered sectors for economic progress and foster innovation will be observed. However, Aslam et al. (2020) proposed the use of the Absolute Innovation Framework to reduce the disruption caused by the introduction of new ideas in a company's operations and combining human intelligence with technical automation for lower product prices and better customer service. The main goal of Industry 5.0 is to foster a harmonious connection between the advanced technology of Industry 4.0 and the human element within the system. Ensuring the safety of workers is of paramount importance in the Industry 5.0 paradigm. According to Maddikunta et al. (2022), Industry 5.0 technology will form the basis of the manufacturing process, leading to higher production, economic growth, and overall prosperity. Ikumapayi et al. (2023) also highlighted the importance of fostering collaboration between humans and robots in the workplace. Human behaviour will play a significant role in shaping the performance of future robots and cobots. Despite the prevalence of misinformation and diverse perspectives, it is undeniable that upcoming technologies, robots, and cobots will undoubtedly progress in terms of their capabilities, intelligence, and productivity. Mass customisation, customer-centric approach, cyber–physical systems, and green computing are key elements that will shape Industry 5.0. These concepts have been extensively studied and discussed by experts in the field (Nahavandi 2019; Skobelev and Borovik 2017; Chander et al. 2022). Duggal et al. (2022) concluded their study with the statement that most systems would achieve full automation by the year 2050. Consequently,

Table 2.2 Comparative highlights of 4.0 and 5.0

Sr. No	Rubrics	Industry 4.0	Industry 5.0
1.	Supply chain	Intelligent	Distributed and responsive
2.	Main focus	Connecting machines	Delivering customers experiences
3.	Manpower	Far from factories	Within factories
4.	Products	Smart	Interactive
5.	Customisation	Mass	Hyper

Table 2.3 Characteristics of Industry 5.0

Industry 5.0	Advanced technologies	AI, ML, IoT, etc. Advanced robotics/cobots Cognitive systems
	Society 5.0	Legal and economic reforms Smart homes and cities Sustainability
	Automation in healthcare	Surgeries with robots/cobots Unified diagnosis Sensors based testing
	Intelligent products, machines, and production	Intelligent and automatic machines Sustainable products Automatic demand based production

users are poised to embark on the upcoming stage of industrial expansion, commonly known as Industry 6.0.

2.5 INDUSTRY 5.0

As stated by many authors (Salimova, Vukovic, and Guskova 2020; ElFar et al. 2021; Duggal et al. 2022; Grabowska, Saniuk, and Gajdzik 2022; Golovianko et al. 2023), the primary goal of Industry 5.0 is to create a system that is human-centric, sustainable, collaborative, and flexible. Table 2.2 (Industry 5.0 2023) provides a comparison of the key features of Industry 4.0 and Industry 5.0. Table 2.3 depicts the characteristics of Industry 5.0.

2.6 TECHNOLOGIES IN INDUSTRY 5.0

The goal of Industry 5.0 is to utilise technology to achieve sustainability and prioritise the well-being of humanity. Various technologies will be utilised in this phase to foster the holistic advancement of humanity and society, with

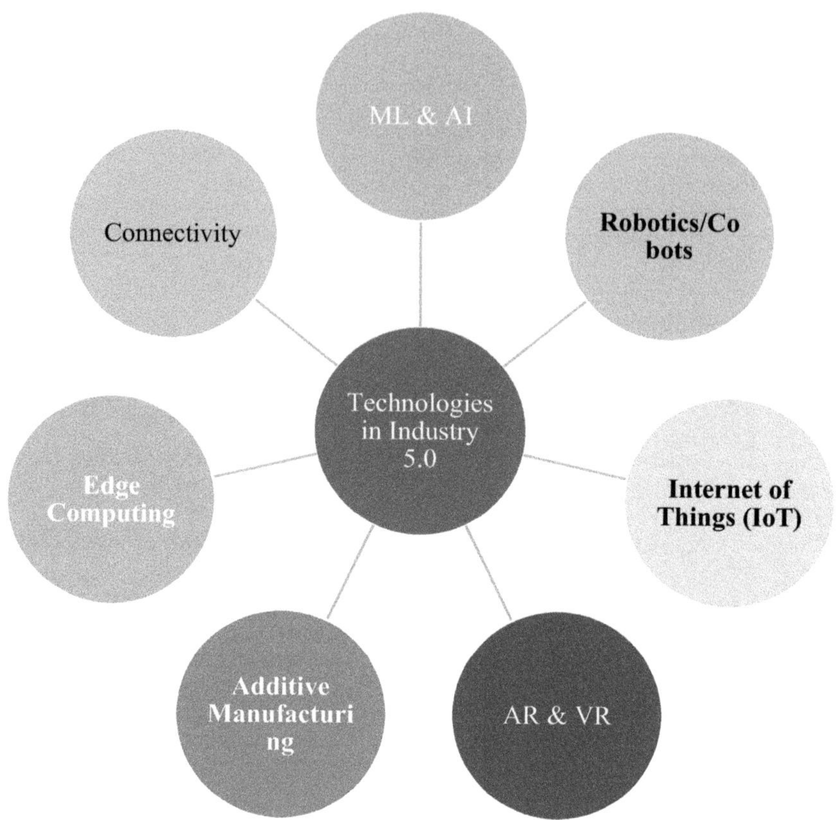

Figure 2.4 Prominent technologies in Industry 5.0.

Source: Own study basis.

a focus on adaptability. Prominent technologies are depicted in Figure 2.4 (Carayannis et al. 2022; Raja Santhi and Muthuswamy 2023):

1. **Machine Learning (ML) and AI:** By leveraging these types of technologies or algorithms, machines can be equipped with AI. ML can greatly enhance the optimisation, prediction, and analysis processes. Having access to this information would greatly benefit industry executives, enabling them to provide informed opinions and forecasts on products, consumer needs, and market trends (Castro et al. 2023) (Garg et al., 2020) (Garg et al., 2023).

2. **Robotics/Cobots:** Recent advancements in robotics and the introduction of collaborative robots (cobots) have opened up new

possibilities for various industries. Manufacturing, healthcare, and other enterprises and organisations can now experience the benefits of humans and robots working together in synergy. It was found that the utilisation of cobots leads to enhancements in accuracy, security, safety, quality, and efficiency (Demir, Döven, and Sezen 2019; Kaasinen et al. 2022).

3. **IoT:** In the current stage, the IoT has become crucial in providing up-to-date and accurate data, which greatly aids in making informed decisions (Chander et al. 2022).

4. **Additive Manufacturing:** It offers the ability to have full control over the supply chain, minimise waste generation, transform mass production, accelerate the prototype process, and produce customised items that perfectly fit your company's requirements.

5. **Edge Computing:** When it comes to mission-critical applications, edge computing outperforms traditional cloud computing by providing rapid response to user interaction and minimising data latency.

6. **AR and VR:** Effective operation of intelligent machines, particularly in the field of AR and VR, relies on proper training for personnel. Thus, AR and VR have the potential to meet the training requirements of the commercial industry. These technological advancements provide highly realistic simulations for teaching and interacting with machines.

7. **Connectivity:** The Internet plays a crucial role in supporting the advanced technologies that drive Industry 5.0. With the advent of 5G technology, numerous benefits have emerged, such as quicker data transfer and access to cloud-based services, enhanced collaboration between professionals and workers, remote machine control, improved data accuracy and efficiency, and more efficient remote machine management.

8. **Cybersecurity:** As collaboration and communication are the main keys in Industry 5.0, data transfer between industries and companies is very common. To prevent cyber frauds and make data or information more secure, strong cybersecurity systems will be applied (Wulandari, Winarno, and Triyanto 2021; Pouyakian 2022).

2.7 OBSTACLES IN IMPLEMENTATION OF INDUSTRY 5.0

In the era of Industry 5.0, there are fascinating speculations surrounding the integration of a diverse range of cutting-edge technologies. With the help of advanced technologies, many well-established industries are set to undergo significant transformations. According to a study conducted by Suryadi, Kushardiyanti, and Gusmanti (2021), it is clear that the present time will present advantages and difficulties for individuals and society. Throughout the implementation process, it is expected that certain challenges may arise (Sharma et al. 2022; Chourasia et al. 2022; Adel 2022; Tavares, Azevedo, and Marques 2022; Lv 2023).

- **Skilled Employees:** With the prevailing dependence on technology in contemporary society, it is imperative to employ personnel who possess a high level of expertise. The future demand for their knowledge remains unknown, particularly in light of the integration of emerging technology. Furthermore, the task of providing training to individuals without prior experience can be a substantial problem for firms.
- **Geographical Challenges:** Geography poses certain challenges due to non-uniform nature of our planet. Challenges related to industry and Internet access are particularly prominent in forested regions, as well as in areas with nearby mountains or hills. It is fascinating how languages can change so quickly within certain countries, like India.
- **Integration and Interoperability:** Integrating pre-existing systems, machines, goods, and raw materials into the design of intelligent machines and products poses a significant technological challenge. Ensuring compatibility between older computers and cutting-edge innovation can pose a challenge. Their approach and utilisation of machinery could potentially exacerbate other challenges.
- **Education:** An informed and knowledgeable population is crucial for the advancement of Industry 5.0. Less knowledgeable countries would face significant challenges, as only skilled individuals can effectively operate advanced robots.
- **Data Security and Privacy:** More data transparency is preferred for equality, yet it may be detrimental to a person or group rather than useful to them. At this point, there is a greater emphasis placed on data-driven machine decision-making. The protection of our personal information and privacy becomes of the utmost importance.
- **Ethical Issues:** The advancement of AI could potentially give rise to a range of new ethical dilemmas. Industrial authorities may encounter a range of challenges, such as determining responsibility in case the machinery disrupts peace.
- **Legal Risk:** With the rise of cutting-edge technologies, it is likely that new legal challenges and uncertainties may arise. To determine the ideal balance between humans and robots, careful consideration must be given to the ethical standards and safeguards that need to be put in place.
- **Electricity/Energy Consumption:** While there is a growing focus on minimising environmental harm, it is inevitable that the usage of

energy and power will rise with the increasing adoption of advanced technology. It is concerning how detrimental this is to the environment, our health, and our survival as it contributes to the acceleration of global warming.

- **Internet and Infrastructure:** The Internet and infrastructure of "Industry 5.0" are backed by advanced digital technologies, autonomous systems, robots and cobots, and state-of-the-art technology. To operate effectively, these technologies require a digital infrastructure and access to high-speed Internet. However, the issue of slow network speeds is a common concern in numerous countries, often influenced by factors such as geographical location or economic conditions.
- **Standardisation:** In a rapidly changing world with diverse individuals, places, and technologies, ensuring consistency across industries can be a challenging task.
- **Implementation:** One of the key considerations is the integration of new technologies with the current workforce and hardware.
- **Collaboration and Communication:** The differences in educational attainment and professional experience across countries and regions, along with the geographical distance between them, pose potential challenges to effective collaboration and communication in the future.
- **Economic Inequalities:** Given the varying gains that certain regions, firms, or industries may experience, it is likely that existing economic disparities will be further exacerbated. Ensuring equal access to technology for all individuals is crucial for addressing inequality and promoting justice.
- **Cultural and Social Obstacles:** There are concerns among certain individuals regarding Industry 5.0, as they express apprehension about potential job loss due to the rise of automation and AI. These individuals may have reservations about placing their trust in these technologies. Due to these biases, achieving a level of recognition and acceptance in society or the workforce that is equal to or higher than 5.0 can be quite challenging.
- **Training:** As the pace of technological advancement continues to accelerate, businesses, industries, and organisations will have to provide retraining opportunities for their current employees to effectively integrate autonomous systems and new technologies. If the purpose of employees is altered in any manner, it will create significant challenges, both from a legal and personal standpoint, for workers and employers.
- **Age:** Another challenge arises from the average age of workers in the workforce. Many individuals in different parts of the world are

approaching retirement age. They seem resistant to adapting to new forms of technology or taking on additional tasks.

- **Cost:** Over this period, the company or organisation will see a rise in its budget due to investments in upgrading machinery, providing worker education, implementing advanced technology, and hiring skilled individuals. Financing collaboration or the development of Industry 5.0 can be more challenging for SMEs compared to large firms.

The focus on human-centricity, cobots, collaboration, and social benefits in Industry 5.0 presents a great opportunity to transform traditional industrial practices. However, implementing Industry 5.0 poses several challenges, as mentioned earlier. To effectively tackle the challenges outlined, it is essential for various stakeholders including businesses, government entities, organisations, academic institutions, and SMEs to collaborate.

2.8 INDUSTRY 5.0 AND SUSTAINABILITY

According to the present study, the achievement of Sustainable Development Goal 9 necessitates the establishment of robust and long-lasting infrastructure, the promotion of equitable and sustainable industrialisation, and the fostering of innovation. Industry 5.0 acknowledges the industry's potential to achieve objectives that extend beyond mere job creation and economic expansion, with a focus on sustainable growth that complies with the environment's limitations and prioritises the well-being of industrial workers throughout the manufacturing process (Solaymani 2019). In spite of their remote geographical positioning, a significant proportion of the global population, exceeding one billion individuals, lack access to essential services, such as transit, as they reside at distances greater than 2 kilometres from the nearest road infrastructure. The transportation industry plays a major role in global carbon dioxide emissions (Romero and Stahre 2021). In 2020, the typical individual's contribution to CO_2 emissions from transportation reached 706 kilograms (Ivanova et al. 2020). The release of greenhouse gases into the atmosphere from different forms of transportation is a significant contributor to the issue of global warming. Governments are faced with the crucial task of improving access to environmentally friendly transport options, as emphasised in the book *"Sustainable Travel for a Liveable Future."* In 2022, the industrial sector was responsible for approximately 25% of the global CO_2 emissions resulting from energy systems, totalling 9.0 Gt of direct emissions. By 2030, the International Energy Agency (IEA,2023) forecasts a notable decrease in industrial emissions to around 7 Gt CO_2. The number does not meet the target set

Table 2.4 Comparison of annual carbon emissions in million tonnes

Years	India	China	Russian Federation	United States of America
1960	32.88	212.9	242.4	788.3
1965	45.26	129.8	317.4	924.7
1970	53.22	210.4	394.7	1181
1975	68.78	312.3	501.7	1202
1980	85.63	400.1	584.1	1288
1985	116.4	536.3	644.5	1225
1990	168.8	671.1	635.2	1315
1995	221.3	905.5	444.9	1401
2000	281.4	928.5	424.8	1555
2005	333.4	1579	440.4	1580
2010	469	2391	456.8	1475
2015	620.7	2788	471.4	1390

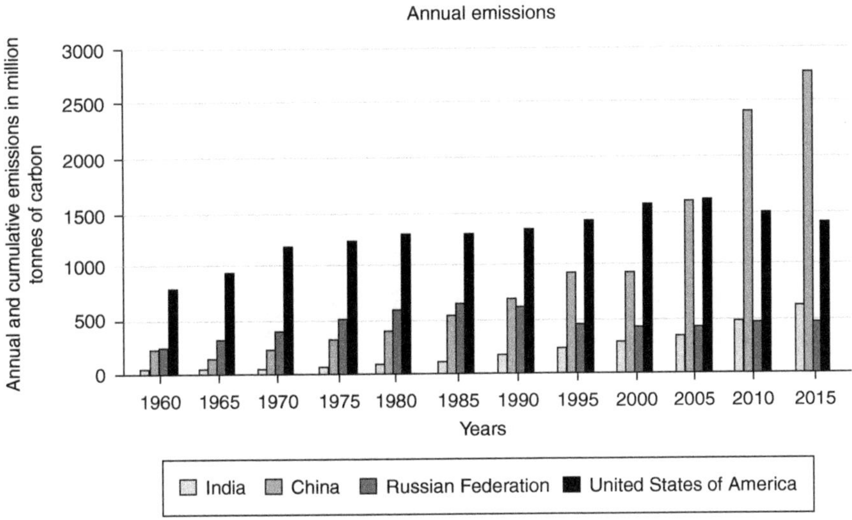

Figure 2.5 Comparison of annual carbon emissions.

Source: Own study basis.

by the Net Zero Emissions by 2050 (NZE) Scenario, which aims to completely eliminate emissions by 2050. There has been a steady increase in the level of CO_2 emissions being released into the atmosphere. Tables 2.4 and 2.5 present a comparison of the annual emissions of India, China, the United States of America, and Russia. The comparative chart of annual emissions and per person emissions of India, China, the United States of America, and Russia is shown in Table 2.4 and Figure 2.5 and Table 2.5

Table 2.5 Comparison of carbon emissions per person in tonnes

Years	India	China	Russian Federation	United States of America
1960	0.07374	0.3255	2.024	4.474
1965	0.0905	0.1793	2.511	4.875
1970	0.0954	0.2558	3.034	5.893
1975	0.1103	0.3414	3.748	5.687
1980	0.1229	0.4073	4.225	5.772
1985	0.1491	0.5058	4.51	5.21
1990	0.194	0.5816	4.292	5.302
1995	0.2295	0.7433	2.995	5.272
2000	0.2656	0.7346	2.893	5.506
2005	0.2887	1.21	3.063	5.324
2010	0.378	1.777	3.189	4.74
2015	0.4692	2.001	3.259	4.282

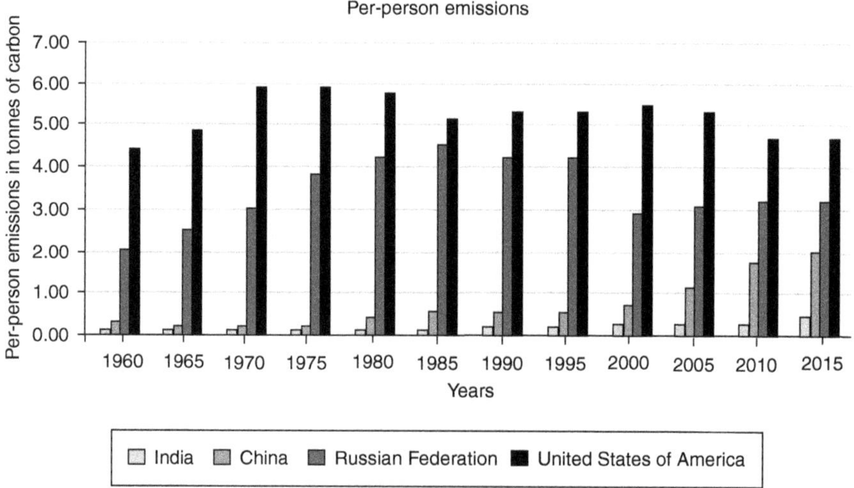

Figure 2.6 Comparison of carbon emissions per person in tonnes.

Source: Own study basis.

and Figure 2.6 (Carbon emissions). So, sustain40ability is a big challenge in the current era. It is expected that technology-based human-centric Industry 5.0 is sustainable for the environment and society (Solaymani 2019; Ivanova et al. 2020).

2.9 ENVIRONMENT PRACTICES WITH SUSTAINABILITY

Industry 5.0, the next phase of industrial development, will play a great role to achieve sustainable goals by 2030. Following are the important factors that will be considered related to sustainability (Shiroishi, Uchiyama, and Suzuki 2018; Contini and Peruzzini 2022; Grabowska, Saniuk, and Gajdzik 2022; Chourasia et al. 2022; Castro et al. 2023).

- **Eco-Friendly Materials:** Industries of the future will prioritise the use of biodegradable materials to create industrial processes that are more environmentally friendly. The waste will undergo biological processes that will transform it into harmless substances, thereby reducing its environmental impact. It will contribute to the reduction of pollution in the air and water, improving waste management, and preserving a pristine natural environment.
- **Decision-Making:** A shift towards a decentralised structure has been determined as a means to mitigate the adverse impact on the environment, thereby transferring decision-making authority from a centralised entity.
- **Circular Economy:** The idea of a "Circular Economy" revolves around an economic theory that suggests minimising resource depletion by reducing waste and instead reprocessing or reusing the material without compromising its attributes. It will contribute to sustaining a robust level of economic activity.
- **Integration of Renewable Energy Sources:** There is a possibility that the upcoming version 5.0 of Industry will incorporate renewable energy sources more extensively and promote their adoption. Embracing renewable energy sources like solar, wind, and hydroelectric power would pave the way for a greener and more sustainable future.
- **Integration of Energy Sources:** With the increasing adoption of technology, there will be significant enhancements in the dependability of energy sources and their interconnections. Enterprises have the potential to shift towards renewable energy sources like solar, wind, and hydroelectric power, leading to a more sustainable future.
- **Smart Cities:** The dream of smart and sustainable cities will be fulfilled with Industry 5.0. Smart cities will be more environmentally friendly through urban design (Kumar et al.,2021). Smart technologies have the potential to reduce building energy usage, improve waste management systems, and generate more environmentally friendly transportation options.
- **Ethical Practices with AI:** In the era of Industry 5.0, as AI and automation continue to advance, the ethical implications of AI practices will gain even more significance. It is essential to carefully

consider the potential consequences for sustainability and the ethical implications of AI-led decisions to minimise any negative impacts on the environment.

- **Sustainability Supply Chain:** Improving traceability and transparency in the supply chain has the potential to reduce the negative effects of raw material extraction and consumption on society and the environment.
- **Recycling and Waste Management:** The improper disposal of waste material, particularly electronic components or appliances, has a profound impact on the sustainability of our environment. Perhaps, the implementation of sustainable recycling methods and the integration of responsible waste management practices could pave the way for a promising future of accelerated technological advancements.
- **Empowerment of Sustainable Innovations:** To accelerate the transition towards a more sustainable economy, the utilisation of cutting-edge technology can assist in the creation and execution of sustainable solutions on a larger scale for business owners.
- **Efficient Resources:** Sophisticated technologies in this phase will improve the optimisation and utilisation of resources. These smart manufacturing processes could be able to reduce waste and energy usage and increase the overall efficiency of processes and resources.
- **Collaborative Sustainability:** In the future, advanced technologies will be utilised to improve resource utilisation and optimisation, leading to more efficient use of available resources. These advanced manufacturing techniques have the potential to greatly decrease the environmental impact of the sector by reducing waste production and energy consumption, while simultaneously enhancing productivity.
- **Green Chemistry:** The course will highlight the significance of implementing eco-friendly chemical practices that make use of renewable resources, with the goal of reducing the environmental impact of manufacturing.
- **Monitoring and Compliance:** Exploring the latest advancements in technology such as ML, the IoT, and AI will allow for ongoing monitoring of the effects of production on people and the environment. These technologies offer up-to-the-minute information on compliance status. These compliance reports will also be useful for quickly addressing the welfare of our communities and the natural ecosystems they depend on.

2.10 IMPLICATIONS OF INDUSTRY 5.0

In the era of Industry 5.0, the significance of automation, robots/cobots, and technological advancements cannot be overstated. Due to its wide-ranging impact on various sectors such as society, industry, education, government, and the stock market, numerous opportunities will arise that were previously inaccessible. Here are some key elements of this stage (Martynov, Shavaleeva, and Zaytseva 2019; Golić 2020; Javaid and Haleem 2020; Cimini et al. 2021; Duggal et al. 2022; Simaiya et al. 2021; Dadwal et al., 2023; Rani et al. 2021; Lilhore et al. 2022a; Adel 2022; Carayannis et al. 2022; Chourasia et al. 2022; Lv 2023):

- **Digital Twins in VR:** The emergence of cobots has led to the advent of digital twins, thereby expanding their potential applications in the realm of VR. This technology will greatly enhance circulation and efficiency in industrial process systems. Professionals with access to advanced visualisation tools can significantly improve production and efficiency by making more informed decisions.
- **Real-Time Systems:** Implementing real-time solutions can significantly boost revenue by closely monitoring production and promptly addressing customer feedback and needs.
- **Simulation and Modelling at Several Scales:** This method of simulation and modelling is highly effective in showcasing the capabilities of the proposed products to potential buyers.
- **Intelligent and Autonomous Systems:** The widespread adoption of intelligent and self-sufficient systems like AI, ML, the IoT, cobots, and others is having a significant impact on every sector of the economy. Implementing these strategies can enhance the autonomy and intelligence of industrial process systems, leading to a positive impact on sustainability.
- **Sensor-Based Data Interoperability:** Interoperability is becoming increasingly important as more and more sensor-based data sources are being used in various settings, including smart homes, cities, autos, workplaces, and other locations. Businesses will discover that various data processes, including those related to sensors, cobots, and other advanced technology, are streamlined, leading to enhanced safety, product quality, and adherence to production schedules.
- **Knowledge Transfer:** As the project progresses, the rules regarding knowledge transfer will gradually disseminate the previous procedures, strategies, security concerns, challenges, and potential advantages.
- **Collaborative Work:** It is crucial to focus on the individuals involved at this stage. As a result, there is a growing anticipation for collaboration between humans and machines. Emphasis should be placed on creativity, originality, and complex problems that can be effectively addressed by robots.
- **AR and Intelligence:** Exploring the potential of AR and intelligence: During this phase, cutting-edge technologies like ML, AI, and other similar advancements will be utilised to enhance human

capabilities rather than solely relying on human cognition. In this approach, humans remain in control of making the most significant decisions, while intelligent robots are solely dedicated to supporting humans in making decisions that align with their best interests.

- **Personalisation:** Encouraging manufacturers to create personalised products based on customer preferences is a key goal of this phase. Personalisation and customisation are essential to prioritise at this stage for various reasons.
- **Ethical Reflections:** The current phase will be more accountable and transparent to avoid bias, and more ethical reflection will be seen in innovation, development, and execution of products with regard to society and humans.
- **Economy and Legal Issues:** As users explore the evolving terrain of Industry 5.0, there will be a growing focus on the economic and legal dimensions. By utilising advanced tools, regulations, and official requirements, government initiatives can attain a higher level of transparency.
- **Healthcare:** There has been a significant transformation in the healthcare industry due to the introduction of robots and cobots. Significant shifts are becoming evident due to the incorporation of IoT, AI, and ML algorithms. Enhanced treatment options can greatly benefit patients by leveraging the power of technology tools.
- **Sustainability:** Understanding and adapting to the current needs and demands are essential for long-term viability. The 17 Sustainable Development Goals outlined by the United Nations are crucial for creating a sustainable society.
- **Productivity:** The adoption of automated systems, machines, robots, intelligent products, and machinery will result in enhanced productivity and decreased errors, costs, and downtime.
- **Safe Industry Environment:** An efficiently organised industry necessitates a meticulously designed infrastructure and a stable environment to facilitate the incorporation of cutting-edge and automated technology. This will lead to a more secure environment for both individuals and businesses.
- **Working Environment:** The atmosphere in which employees perform their duties can greatly affect the functioning of companies and other organisations. Experienced professionals will be assigned to different shifts, working alongside cutting-edge equipment.
- **Global Growth:** The expected increase in industry collaboration is poised to have a significant worldwide influence. The collaboration of various parties, businesses worldwide will flourish, resulting in a surge of job opportunities.

2.11 CONCLUSIONS

The integration of intelligent systems across industries will have a significant impact on the conception, production, and distribution of products in the era of the Internet. The greatest advantage of Industry 5.0 lies in its ability to enhance production efficiency and productivity. Intelligent systems optimise the use of resources, minimise downtime, and enhance productivity through the analysis of real-time data, anticipation of maintenance needs, and adaptable manufacturing processes. Intelligent systems also contribute to cost savings for businesses. Enhancing manufacturing operational efficiency can greatly improve competitiveness and sustainability, resulting in significant cost savings. Product quality is enhanced through intelligent manufacturing in the Fifth Industrial Revolution. By utilising advanced technology, the implementation of intelligent systems in manufacturing processes helps minimise errors and ensures consistent product quality. In today's market, maintaining good quality is crucial for organisations to consistently deliver on their promises of accuracy and reliability, meeting the expectations of customers. Manufacturers now have the ability to customise their products to meet the specific requirements of their clients. The Industry 5.0 revolution places a strong emphasis on minimising waste and maximising efficiency. The industrial sector, notorious for its environmental impact, is expected to adopt a more environmentally conscious approach. Implementing energy-efficient practices, reducing waste, and adopting environmentally sensitive production methods can contribute to a more sustainable and ecologically friendly manufacturing process. Therefore, the needs of environmentally conscious consumers are met while also supporting the global environmental movement.

Smart manufacturing also highlights the importance of collaboration between humans and machines. Enhancing supply chain operations could be achieved through the seamless integration of intelligence into industrial processes. By enhancing supply chain efficiency and reducing costs, customer service and financial gains are improved (Voulgaridis et al.,2022). There are certain obstacles that need to be overcome on the journey towards Industry 5.0. Considering the importance of gathering and examining data, it is essential to prioritise data privacy and confidentiality. Ensuring the security of production data is crucial in the realm of smart manufacturing. It is crucial for all manufacturers to place a high emphasis on implementing rigorous safety protocols and robust data protection measures. As this industry continues to expand, it will require additional regulatory and ethical measures. This event will explore the ethical considerations surrounding robots, the importance of protecting intellectual property, and the need for robust information security measures. Manufacturers need to embrace change, foster innovation, and adjust to smart manufacturing to stay competitive in today's rapidly evolving industrial landscape.

REFERENCES

Adel, A. (2022). Future of Industry 5.0 in society: Human-centric solutions, challenges and prospective research areas. *Journal of Cloud Computing, 11*(1). https://doi.org/10.1186/s13677-022-00314-5

Akundi, A., Euresti, D., Luna, S., Ankobiah, W., Lopes, A., & Edinbarough, I. (2022). State of Industry 5.0—analysis and identification of current research trends. *Applied System Innovation, 5*(1), 1–14. https://doi.org/10.3390/asi5010027

Aslam, F., Aimin, W., Li, M., & Rehman, K. U. (2020). Innovation in the era of IoT and Industry 5.0: Absolute innovation management (AIM) framework. *Information (Switzerland), 11*(2). https://doi.org/10.3390/info11020124

Brunetti, D., Gena, C., & Vernero, F. (2022). Smart interactive technologies in the human-centric factory 5.0: A survey. *Applied Sciences (Switzerland), 12*(16). https://doi.org/10.3390/app12167965

Carayannis, E. G., Dezi, L., Gregori, G., & Calo, E. (2022). Smart environments and techno-centric and human-centric innovations for industry and Society 5.0: A quintuple helix innovation system view towards smart, sustainable, and inclusive solutions. *Journal of the Knowledge Economy, 13*(2), 926–955. https://doi.org/10.1007/s13132-021-00763-4

Carbon emissions: China+united-states-of-america+russian-federation. *Population Pyramid.net.* (n.d.). www.populationpyramid.net/carbon-emissions/china+united-states-of-america+russian-federation/

Castro, H., Costa, F., Ferreira, T., Ávila, P., Cruz-Cunha, M., Ferreira, L., Putnik, G. D., & Bastos, J. (2023). Data science for Industry 4.0 and sustainability: A survey and analysis based on open data. *Machines, 11*(4). https://doi.org/10.3390/machines11040452

Chaka, C. (2022). Is Education 4.0 a sufficient innovative, and disruptive educational trend to promote sustainable open education for higher education institutions? A review of literature trends. *Frontiers in Education, 7*(April), 1–13. https://doi.org/10.3389/feduc.2022.824976

Chander, B., Pal, S., De, D., & Buyya, R. (2022). Artificial intelligence-based Internet of Things for Industry 5.0. In *Artificial Intelligence-Based Internet of Things Systems*, 3–45. https://doi.org/10.1007/978-3-030-87059-1_1

Chourasia, S., Tyagi, A., Pandey, S. M., Walia, R. S., & Murtaza, Q. (2022). Sustainability of Industry 6.0 in global perspective: Benefits and challenges. *MAPAN—Journal of Metrology Society of India, 37*(2), 443–452. https://doi.org/10.1007/s12647-022-00541-w

Cimini, C., Boffelli, A., Lagorio, A., Kalchschmidt, M., & Pinto, R. (2021). How do Industry 4.0 technologies influence organisational change? An empirical analysis of Italian SMEs. *Journal of Manufacturing Technology Management, 32*(3), 695–721. https://doi.org/10.1108/JMTM-04-2019-0135

Contini, G., & Peruzzini, M. (2022). Sustainability and Industry 4.0: Definition of a set of key performance indicators for manufacturing companies. *Sustainability (Switzerland), 14*(17). https://doi.org/10.3390/su141711004

Dadwal, S., Kumar, P., Verma, R., & Singh, G. (n.d.). *Opportunities and Challenges of Business 5.0 in Emerging Markets*. IGI Global, 1–357. https://doi.org/10.4018/978-1-6684-6403-8

Dar Masroof Amin, A. G. (2014). A proposed framework using neural network in web mining for improving the performance of e-learning system. *International Journal of Science and Research (IJSR)*, 12(12), 2681–2688. www.ijsr.net/get_abstract.php?paper_id=SUB141145

Demir, K. A., & Cicibaş, H. (2019). The next industrial revolution: Industry 5.0 and discussions on Industry 4.0. In: *4th International Management Information Systems Conference "Industry* (Vol. 4, pp. 17–20).

Demir, K. A., Döven, G., & Sezen, B. (2019). Industry 5.0 and human–robot co-working. *Procedia Computer Science, 158*, 688–695. https://doi.org/10.1016/j.procs.2019.09.104

Duggal, A. S., Malik, P. K., Gehlot, A., Singh, R., Gaba, G. S., Masud, M., & Al-Amri, J. F. (2022). A sequential roadmap to Industry 6.0: Exploring future manufacturing trends. *IET Communications, 16*(5), 521–531. https://doi.org/10.1049/CMU2.12284

Elbeltagi, A., Aslam, M. R., Malik, A., Mehdinejadiani, B., Srivastava, A., Bhatia, A. S., & Deng, J. (2020). The impact of climate changes on the water footprint of wheat and maize production in the Nile Delta, Egypt. *Science of the Total Environment, 743*, 140770. https://doi.org/10.1016/j.scitotenv.2020.140770

ElFar, O. A., Chang, C. K., Leong, H. Y., Peter, A. P., Chew, K. W., & Show, P. L. (2021). Prospects of Industry 5.0 in algae: Customization of production and new advance technology for clean bioenergy generation. *Energy Conversion and Management: X, 10*(April 2020), 100048. https://doi.org/10.1016/j.ecmx.2020.100048

Emma-Ikata, D., & Doyle-Kent, M. (2022). Industry 5.0 readiness—"optimization of the relationship between humans and robots in manufacturing companies in southeast of Ireland." *IFAC-PapersOnLine, 55*(39), 419–424. https://doi.org/10.1016/j.ifacol.2022.12.071

Garg, A., Sharma, A., & Garg, N. B. (2020). Impact of web based learning and teaching in higher education in India. *Journal of Computational and Theoretical Nanoscience, 17*(6), 2689–2694. https://doi.org/10.1166/jctn.2020.8968

Garg, A., Garg, N. B., Gupta, K., & Prasad, D. (2021a). Impact of computer applications to study mechanical properties of HSLA steels. In: *2021 International Conference on Computational Intelligence and Computing Applications (ICCICA)*. IEEE, 1–5. https://doi.org/10.1109/ICCICA52458.2021.9697181

Garg, A., Lilhore, U. K., Ghosh, P., Prasad, D., & Simaiya, S. (2021b). Machine learning-based model for prediction of student's performance in higher education. In: *2021 8th International Conference on Signal Processing and Integrated Networks (SPIN)*. IEEE, 162–168. https://doi.org/10.1109/spin52536.2021.9565999

Garg, A., Garg, N., Lilhore, U., Popli, R., Simaiya, S., & Bansal, A. (2023). Machine learning-based model to predict student's success in higher education. In: *Proceedings of the 3rd International Conference on ICT for Digital, Smart, and Sustainable Development, ICIDSSD 2022, 24–25 March 2022, New Delhi, India* (p. 17). European Alliance for Innovation. https://doi.org/10.4108/eai.24-3-2022.2318766

Gholami, H., Abu, F., Lee, J. K. Y., Karganroudi, S. S., & Sharif, S. (2021). Sustainable manufacturing 4.0—pathways and practices. *Sustainability (Switzerland)*, *13*(24). https://doi.org/10.3390/SU132413956

Golić, Z. (2020). Finance and artificial intelligence: The fifth industrial revolution and its impact on the financial sector. *Зборник Радова Економског Факултета У Источном Сарајеву*, *8*(19), 67. https://doi.org/10.7251/zrefis1919067g

Golovianko, M., Terziyan, V., Branytskyi, V., & Malyk, D. (2023). Industry 4.0 vs. Industry 5.0: Co-existence, transition, or a hybrid. *Procedia Computer Science*, *217*(2022), 102–113. https://doi.org/10.1016/j.procs.2022.12.206

Grabowska, S., Saniuk, S., & Gajdzik, B. (2022). Industry 5.0: Improving humanization and sustainability of Industry 4.0. *Scientometrics*, *127*(6), 3117–3144. https://doi.org/10.1007/s11192-022-04370-1

Huang, S., Wang, B., Li, X., Zheng, P., Mourtzis, D., & Wang, L. (2022). Industry 5.0 and Society 5.0—comparison, complementation and co-evolution. *Journal of Manufacturing Systems*, *64*, 424–428. https://doi.org/10.1016/j.jmsy.2022.07.010

IEA (2023). *Greenhouse Gas Emissions from Energy Data Explorer*. IEA, Paris. www.iea.org/data-and-statistics/data-tools/greenhouse-gas-emissions-from-energy-data-explorer

Ikumapayi, O. M., Afolalu, S. A., Ogedengbe, T. S., Kazeem, R. A., & Akinlabi, E. T. (2023). Human–robot co-working improvement via revolutionary automation and robotic technologies—an overview. *Procedia Computer Science*, *217*, 1345–1353. https://doi.org/10.1016/j.procs.2022.12.332

Industry 5.0 – the essence and reasons why it gets more attention (2024) i. Available at: www.i-scoop.eu/industry-4-0/industry-5-0/ (Accessed: 29 Nov 2023)

Ingaldi, M., & Ulewicz, R. (2020). Problems with the implementation of Industry 4.0 in enterprises from the SME sector. *Sustainability (Switzerland)*, *12*(1). https://doi.org/10.3390/SU12010217

Jafari, N., Azarian, M., & Yu, H. (2022). Moving from Industry 4.0 to Industry 5.0: What are the implications for smart logistics? *Logistics*, *6*(2), 1–27. https://doi.org/10.3390/logistics6020026

Javaid, M., & Haleem, A. (2020). Critical components of Industry 5.0 towards a successful adoption in the field of manufacturing. *Journal of Industrial Integration and Management*, *5*(3), 327–348. https://doi.org/10.1142/S2424862220500141

Kaasinen, E., Anttila, A. H., Heikkilä, P., Laarni, J., Koskinen, H., & Väätänen, A. (2022). Smooth and resilient human–machine teamwork as an Industry 5.0 design challenge. *Sustainability (Switzerland)*, *14*(5). https://doi.org/10.3390/SU14052773

Klimuk, V., Tarasova, A., Yulia, K., & Laura, D. (2020). Synergistic interaction of education, science, and industry. *Leadership, Education, Personality: An Interdisciplinary Journal*, *2*(1), 53–58. https://doi.org/10.1365/s42681-020-00009-y

Ko, M., Kim, C., Lee, S., & Cho, Y. (2020). An assessment of smart factories in Korea: An exploratory empirical investigation. *Applied Sciences (Switzerland)*, *10*(21), 1–15. https://doi.org/10.3390/app10217486

Kumar, A., Sharma, S., Goyal, N., Singh, A., Cheng, X., & Singh, P. (2021). Secure and energy-efficient smart building architecture with emerging technology IoT. *Computer Communications*, *176*(June), 207–217. https://doi.org/10.1016/j.comcom.2021.06.003

Lilhore, U. K., Simaiya, S., Pandey, H., Gautam, V., Garg, A., & Ghosh, P. (2022a). Breast cancer detection in the IoT cloud-based healthcare environment using fuzzy cluster segmentation and SVM classifier. *Lecture Notes in Networks and Systems*, *356*, 165–179. https://doi.org/10.1007/978-981-16-7952-0_16

Lilhore, U. K., Simaiya, S., Sandhu, J. K., Trivedi, N. K., Garg, A., & Moudgil, A. (2022b). Deep learning-based predictive model for defect detection and classification in Industry 4.0. In: *2022 International Conference on Emerging Smart Computing and Informatics (ESCI)*. IEEE, 1–5. https://doi.org/10.1109/ESCI53509.2022.9758280

Lv, Z. (2023). Digital twins in Industry 5.0. *Research*, *6*, 1–16. https://doi.org/10.34133/research.0071

Maddikunta, P. K. R., Pham, Q.-V., Prabhadevi, B., Deepa, N., Dev, K., Gadekallu, T. R., Ruby, R., & Liyanage, M. (2022). Industry 5.0: A survey on enabling technologies and potential applications. *Journal of Industrial Information Integration*, *26*. https://doi.org/10.1016/j.jii.2021.100257

Martynov, V. V., Shavaleeva, D. N., & Zaytseva, A. A. (2019). Information technology as the basis for transformation into a digital society and Industry 5.0. In: Proceedings of the 2019 IEEE International Conference Quality Management, Transport and Information Security, Information Technologies *(IT&QM&IS)*. IEEE, 539–543. https://doi.org/10.1109/ITQMIS.2019.8928305

Moeuf, A., Pellerin, R., Lamouri, S., Tamayo-Giraldo, S., & Barbaray, R. (2018). The industrial management of SMEs in the era of Industry 4.0. *International Journal of Production Research*, *56*(3), 1118–1136. https://doi.org/10.1080/00207543.2017.1372647

Nahavandi, S. (2019). Industry 5.0—a human-centric solution. *Sustainability (Switzerland)*, *11*(16). https://doi.org/10.3390/SU11164371

Nurmila, N., Supiana, R. I., & Aripudin, A. (2021). The concept of gender education in the family to help achieve sustainable development goals. In: *Proceedings of the 5th Asian Education Symposium 2020 (AES 2020)*. Atlantis Press, *566*(AES 2020), 427–431. https://doi.org/10.2991/assehr.k.210715.089

Özdemir, V., & Hekim, N. (2018). Birth of Industry 5.0: Making sense of Big Data with artificial intelligence, "the Internet of Things" and next-generation technology policy. *OMICS—A Journal of Integrative Biology*, *22*(1), 65–76. https://doi.org/10.1089/OMI.2017.0194

Pilevari, N., & Yavari, F. (2020). Journal of industrial strategic management industry revolutions development from Industry 1.0 to Industry 5.0 in manufacturing industry revolutions development from Industry 1.0 to Industry 5.0 in manufacturing. *Journal of Industrial Strategic Management*, *2020*(2), 44–63.

Pouyakian, M. (2022). Cybergonomics: Proposing and justification of a new name for the ergonomics of Industry 4.0 technologies. *Frontiers in Public Health*, *10*. https://doi.org/10.3389/fpubh.2022.1012985

Rani, S., Koundal, D., Kavita, Ijaz, M. F., Elhoseny, M., & Alghamdi, M. I. (2021). An optimized framework for WSN routing in the context of Industry 4.0. *Sensors (Basel, Switzerland)*, *21*(19), 1–15. https://doi.org/10.3390/s21196474

Romero, D., & Stahre, J. (2021). Towards the resilient Operator 5.0: The future of work in smart resilient manufacturing systems. *Procedia CIRP*, *104*(March), 1089–1094. https://doi.org/10.1016/j.procir.2021.11.183

Salimova, T., Vukovic, N., & Guskova, N. (2020). Towards sustainability through Industry 4.0 and Society 5.0. *International Review*, *3–4*, 48–54. https://doi.org/10.5937/INTREV2003048S

Saxena, A., Saxena, A., Sharma, R., & Parashar, M. (2021). Emergence of futuristic HRM in perspective of human—cobot's collaborative functionality. *International Journal of Engineering and Advanced Technology*, *10*(5), 292–296. https://doi.org/10.35940/ijeat.e2763.0610521

Sharma, A., & Singh, B. J. (2020). Evolution of industrial revolutions: A review. *International Journal of Innovative Technology and Exploring Engineering*, *9*(11), 66–73. https://doi.org/10.35940/ijitee.i7144.0991120

Sharma, M., Sehrawat, R., Luthra, S., Daim, T., & Bakry, D. (2022). Moving towards Industry 5.0 in the pharmaceutical manufacturing sector: Challenges and solutions for Germany. *IEEE Transactions on Engineering Management*, 1–18. https://doi.org/10.1109/TEM.2022.3143466

Shiroishi, Y., Uchiyama, K., & Suzuki, N. (2018). Society 5.0: For human security and well-being. *Computer*, *51*(7), 91–95. https://doi.org/10.1109/MC.2018.3011041

Sinha, S., Sharma, S., Johari, S., Sharma, A., & Rajkhowa, S. (2023). Design and development of hydrophobicity and net charge based artificial neural network model for IDP/IDPR prediction. *Procedia Computer Science*, *218*(2022), 438–448. https://doi.org/10.1016/j.procs.2023.01.026

Skobelev, P., & Borovik, Y. (2017). On the way from Industry 4 .0 to Industry 5.0. *International Scientific Journal "Industry 4.0,"* *2*(6), 307–311. https://stumejournals.com/journals/i4/2017/6/307/pdf

Sołtysik-Piorunkiewicz, A., & Zdonek, I. (2021). How Society 5.0 and Industry 4.0 ideas shape the open data performance expectancy. *Sustainability (Switzerland)*, *13*(2), 1–24. https://doi.org/10.3390/su13020917

Solaymani, S. (2019). CO2 emissions patterns in 7 top carbon emitter economies: The case of transport sector. *Energy*, *168*, 989–1001.

Suryadi, S., Kushardiyanti, D., & Gusmanti, R. (2021). Challenges of community empowerment in the era of industry Society 5.0. *KOLOKIUM Jurnal Pendidikan Luar Sekolah*, *9*(2), 160–176. https://doi.org/10.24036/kolokium-pls.v9i2.492

Tavares, M. C., Azevedo, G., & Marques, R. P. (2022). The challenges and opportunities of era 5.0 for a more humanistic and sustainable society—a literature review. *Societies*, *12*(6), 1–21. https://doi.org/10.3390/soc12060149

Turner, C., Oyekan, J., Garn, W., Duggan, C., & Abdou, K. (2022). Industry 5.0 and the circular economy: Utilizing LCA with intelligent products. *Sustainability (Switzerland)*, *14*(22). https://doi.org/10.3390/su142214847

Vaidya, S., Ambad, P., & Bhosle, S. (2018). Industry 4.0—a glimpse. *Procedia Manufacturing*, *20*, 233–238. https://doi.org/10.1016/j.promfg.2018.02.034

Voulgaridis, K., Lagkas, T., & Sarigiannidis, P. (2022). Towards Industry 5.0 and digital circular economy: Current research and application trends. In: 2022 18th Annual International Conference on Distributed Computing in Sensor Systems (DCOSS). IEEE, 153–158. https://doi.org/10.1109/DCOSS54 816.2022.00037

Wang, B., Tao, F., Fang, X., Liu, C., Liu, Y., & Freiheit, T. (2021). Smart manufacturing and intelligent manufacturing: A comparative review. *Engineering*, 7(6), 738–757. https://doi.org/10.1016/j.eng.2020.07.017

Wulandari, E., Winarno, W., & Triyanto, T. (2021). Digital citizenship education: Shaping digital ethics in Society 5.0. *Universal Journal of Educational Research*, 9(5), 948–956. https://doi.org/10.13189/ujer.2021.090507

Xu, X., Lu, Y., Vogel-Heuser, B., & Wang, L. (2021). Industry 4.0 and Industry 5.0— inception, conception and perception. *Journal of Manufacturing Systems*, 61, 530–535. https://doi.org/10.1016/j.jmsy.2021.10.006

Chapter 3

Customer-centric stock broking in Industry 5.0

An examination of awareness, perception, and expectation for satisfaction

Himanshu Lad, Aparna Vajpayee, and Parag Sanghani

3.1 INTRODUCTION

In the context of Industry 5.0, stock broking firms serve as pivotal advisors to individuals seeking investment and wealth management solutions. These firms offer a diverse range of products and services facilitated by cutting-edge technological tools, fostering an environment of enhanced investor participation in the equity market. Leveraging the transformative capabilities of Industry 5.0, investors now engage in a multitude of trading activities encompassing delivery-based trading, intraday transactions, futures and options, commodity trading, and currency exchange. The revenue models of stock broking firms primarily hinge on brokerage charges per trading endeavour. Notable industry players, including Axis Direct, Motilal Oswal, Sharekhan, and others, vie for investors' allegiance within a competitive landscape, with customer-centric factors like branch accessibility, facilities quality, technological infrastructure, and transactional security exerting considerable influence over customer retention. Consequently, nurturing customer satisfaction becomes a pivotal mission for these firms, necessitating periodic evaluations of client contentment across the spectrum of services provided.

3.2 LITERATURE REVIEW

Kajapriya and Venkateswaran (2021) found that performance of Indian Stock Market is highly influenced by the stock broking companies. The stock broking firms are continuously making changes in their business model. They mainly focus on the customer perception and expectations towards the services provided by the stock broking firm.

Agrawal and Mittal (2019) found that customer satisfaction can be improved by customer trust and involvement in Customer relationship management (CRM) system. CRM system also influences customer retention and customer loyalty. The article will help stock broking firms with their

DOI: 10.1201/9781032677040-3

tactical decision for CRM implementation. CRM system must be measuring effectiveness of stock broking firm from the customer perspective.

Meenakshi (2018) found that study of customer satisfaction of a particular product or service will knowledge about the company's performance is in the market. If the satisfaction of the customer increases, then it results in profitability of the company. So, it becomes necessary for enterprises to study the satisfaction factors and levels Trivedi, Mehata and Sharma (2021) investigated the role of customer service quality in customer satisfaction within the stock broking industry. Their study found that customers who perceive higher service quality, including prompt response to queries, knowledgeable staff, and efficient trade execution, tend to have higher levels of satisfaction. The findings emphasize the importance of delivering excellent customer service as a means to enhance overall customer satisfaction (Yang, and Fang 2004).

Gupta and Verma (2020) explored the impact of technological advancements on customer satisfaction in the context of online trading platforms. Their research revealed that customers who perceive online trading platforms to be user-friendly, secure, and equipped with advanced features are more likely to be satisfied with the services. The study highlights the significance of incorporating user-friendly technology and ensuring a seamless online trading experience to foster customer satisfaction.

Lee and Park (2019) conducted a study on the role of trust in customer satisfaction within the stock broking industry. Their research indicated that customers who have a high level of trust in their stock broking firms, including trust in the accuracy of information provided and the confidentiality of personal data, tend to exhibit higher levels of satisfaction. The study emphasizes the importance of building and maintaining trust-based relationships with customers to enhance their satisfaction levels (Rajkumar and Rangaswami 2017).

Chen and Chen (2018) examined the influence of personalized services on customer satisfaction in stock broking firms. Their findings revealed that customers who receive personalized services, such as tailored investment advice, customized portfolio management, and personalized communication, tend to have higher levels of satisfaction. The study underscores the significance of offering personalized services that meet individual customer needs and preferences to enhance overall satisfaction.

Gupta and Agarwal (2017) conducted a study on the impact of transparency in pricing and fee structures on customer satisfaction in the stock broking industry. Their research indicated that customers who perceive transparency in pricing, including clear information on brokerage charges, account maintenance fees, and transaction costs, tend to have higher levels of satisfaction. The study highlights the importance of transparent pricing practices as a driver of customer satisfaction.

Wang and Li (2016) examined the role of customer engagement in customer satisfaction within the stock broking industry. Their research revealed that customers who are actively engaged with their stock broking firms, participating in educational seminars, accessing research reports, and engaging in community forums, tend to exhibit higher levels of satisfaction. The study emphasizes the importance of fostering customer engagement through various channels to enhance overall customer satisfaction (Sharma and Mehta, 2023).

These additional literature reviews provide further insights into various factors that influence customer satisfaction in the stock broking industry. They highlight the significance of customer service quality, technological advancements, trust, personalized services, transparency in pricing, and customer engagement as key drivers of customer satisfaction. By considering these factors, stock broking firms can align their strategies and practices to meet customer expectations and enhance overall satisfaction levels (Jeyaprabha and Sunder 2022.

3.3 SIGNIFICANCE OF THE RESEARCH IN THE CONTEXT OF INDUSTRY 5.0

Stock broking companies stand as critical intermediaries between investors and listed companies within the framework of Industry 5.0. Recognizing their profound influence on market dynamics becomes imperative for regulators, policymakers, and market participants, especially in the context of the Industry 5.0 transformation. With the Indian stock market occupying a prominent position as a dynamic and expansive market, investigating the role of stock broking companies in shaping its performance yields insights into the functioning of emerging market economies. Furthermore, this research contributes to the evolving understanding of financial markets within the Industry 5.0 paradigm.

Customer satisfaction and trust remain paramount for the success of stock broking companies, especially as Industry 5.0 redefines service standards. Delving into the interplay between customer satisfaction and market performance, this research unravels the factors driving growth and stability in the stock market. It serves to illuminate the efficacy of prevailing regulations and market practices governing stock broking companies. The findings can pinpoint areas necessitating improvements or regulatory adjustments, ensuring equity and efficiency in an evolving market landscape shaped by Industry 5.0.

As Industry 5.0 catalyses transformative technological progress within the financial sector, comprehending the repercussions of these advancements on customer satisfaction and market performance gains heightened significance. The research investigates how technology dovetails with the operations of stock broking companies to elevate performance and its broader implications for the market ecosystem. By addressing these research aims, the study seeks to unravel the intricate relationships among stock broking entities, customer contentment, and the performance of the Indian stock market within the Industry 5.0 framework. The research's findings stand to steer market stakeholders, regulators, and policymakers toward strategies fostering transparency, efficiency, and investor trust in an Industry 5.0-driven financial landscape.

3.4 RESEARCH OBJECTIVES WITHIN THE INDUSTRY 5.0 FRAMEWORK

The research encapsulates the following objectives within the Industry 5.0 context:

i. **Analysing Stock Broking Companies:** Impact on the Indian Stock Market Performance in the Industry 5.0 Era: This research probes the extent of stock broking companies' influence on the overall performance and stability of the Indian stock market, unravelling their role within the Industry 5.0 ecosystem.

ii. **Exploring the Nexus between Customer Satisfaction and Stock Broking Company Performance:** Investigating the correlation between customer satisfaction levels and stock broking company performance, the research dissects whether elevated customer satisfaction translates into improved financial outcomes and market dynamics within Industry 5.0.

iii. **Evaluating Technological Advancements: Effects on Customer Satisfaction and Market Performance:** In the Industry 5.0 landscape, the study assesses the impact of technological advancements like digital platforms on customer satisfaction and the comprehensive performance of stock broking companies, unveiling insights into the reshaping dynamics of the sector.

iv. **Identifying Customer Satisfaction Catalysts in the Stock Broking Industry:** This research identifies and scrutinizes the key drivers behind customer satisfaction in the stock broking industry within the Industry 5.0 paradigm. Such insights inform refined service provision and superior customer experiences.

v. **Proposing Recommendations for Augmenting Customer Satisfaction and Market Performance:** With findings as a foundation, this research aims to offer pragmatic recommendations and strategies for stock broking companies, regulators, and policymakers, with the intention of elevating customer satisfaction and ensuring the efficiency and stability of the Indian stock market in the Industry 5.0 era.

The culmination of these objectives paves the way for a deeper comprehension of the interplay among stock broking entities, customer contentment, and the performance of the Indian stock market within the context of Industry 5.0. The research's outcomes stand poised to guide industry practices, regulatory decisions, and strategic pursuits towards fostering a resilient, Industry 5.0-aligned stock market environment driven by customer-centric principles.

3.5 RESEARCH METHODOLOGY

The objective of the study is to identify customer awareness, perception, attitude, and satisfaction towards stock broking firms in Surat city. The researchers started collection of data from December 2022 to March 2023. The sample frame of the study was Surat city. We used non-probability convenience sampling method to collect samples. We used an online structured questionnaire to collect the data from 142 respondents of Surat city. We also used books, websites, and magazines as secondary data. The SPSS software was used for data analysis. In this paper, we applied factor analysis, chi-square test, and correlation for getting relations between the different variables.

3.6 DATA ANALYSIS

We collected data from 142 respondents of which 138 are male and only 4 respondents are female. Sixty percent of respondents are from the age group 31 to 40 years, 21% of respondents are from 41 to 50 years, 13% of respondents are from 18 to 30 years, and only 6% of respondents are above 51 years. Ninety percent of respondents are graduate, 6% of respondents are post graduate, and only 4% of respondents are undergraduate. Thirty-five percent of respondents are business persons, 28% of respondents are salaried employees, and very few are students and retired.

3.6.1 Hypothesis I

H0: There is no significant association between awareness of future trading options with demographic variables.
H1: There is a significant association between awareness of future trading options with demographic variables.

The p value of the chi-square test is shown in Table 3.1, which indicates that there is a significant association between awareness of future trading options and demographic factors, age, gender, income, and education.

Table 3.1 Cross-Tabulation between awareness about future trading options and demographic

Cross Tabulation	Demographic	Pearson's Sig. Value	Relation
Awareness about	Age	0.026	Exist
Future Trading	Gender	0.010	Exist
Options	Income	0.000	Exist
	Education	0.017	Exist
	Occupation	0.342	Absent

Table 3.2 KMO and Bartlett's test

Kaiser–Meyer–Olkin Measure of Sampling Adequacy		0.835
Bartlett's test of sphericity	Approx. chi-square	3348.158
	df	210
	Sig.	0.000

3.6.2 Hypothesis 2: Factor analysis

H0: Correlation matrix is an identity matrix.
H1: Correlation matrix is not an identity matrix (desirable).

As shown in Table 3.2, the Kaiser-Meyer-Olkin (KMO) and Bartlett's test have a significant value of 0.000, which is less than 0.05. So, we reject the null hypothesis and prove that the correlation matrix is not an identity matrix. Therefore, we bifurcate the statements into four groups. As shown in Table 3.3, the first component shows the group of Trading Tips, the second component shows the group of Branch Facility, the third component shows the group of Customer Handling, and the fourth component shows the group of Account Handling.

3.6.3 Hypothesis 3: Regression analysis

H0: All the independent factors do not affect the dependent variable expectation of return.
H1: All the independent factors affect the dependent variable expectation of return (Table 3.4).

As shown in Table 3.5, the regression analysis test significant value is 0.000, which is less than 0.05. So, reject H0 and prove that all independent factors affect the dependent variable. As shown in Table 3.6, in the normality test, the significant value of Kolmogorov–Smirnov and Shapiro–Wilk is 0.000, which is also valid for the regression equation.

Table 3.3 Rotated component matrix

	Component			
	1	*2*	*3*	*4*
Nifty & Bank nifty; Hero or zero; call is profitable on Thursday.	0.867			
Fin nifty Hero or Zero call is profitable on Tuesday.	0.860			
Currency segment tips are helpful for profitable trades.	0.858			
Nifty & Bank nifty options tips are helpful for profitable trades	0.818			
Equity cash tips are helpful for profitable trades.	0.786			
Nifty & Bank nifty levels are helpful for profitable trades.	0.786			
Branch location is convenient.		0.880		
External appearance of branch is attractive.		0.873		
Good terminal facility for customers.		0.856		
Branch has clean environment		0.811		
Branch has good infrastructure facilities like TV, sofa, chairs, AC, etc.		0.810		
Good ability to explain trading procedures.			0.884	
Good ability to explain the brokerage charges and other charges.			0.863	
Good ability to help purchase or choose the right shares at the right time.			0.861	
Good ability to explain procedures to open demat a/c.			0.856	
Good ability for clarifying customer doubts regarding demat a/c.			0.831	
Procedure of transferring the money from trading a/c to saving bank a/c and vice versa is easy.				0.863
Transactions of trading is accurate and easy.				0.848
KYC (Know Your Customer) updating facility is easy.				0.826
Demat a/c opening procedure is easy.				0.805
Portfolio management services is easy.				0.784

Table 3.4 Model summary

Model	R	R Square	Adjusted R Square	Std. Error of the Estimate
1	0.390[a]	0.152	0.127	0.651

a. Predictors: (Constant), Account Handling, Customer Handling, Branch Facility, Trading Tips

b. Dependent Variable: What is the rate of return expected by you from Equity Market in a year?

Table 3.5 ANOVA

Model		Sum of Squares	df	Mean Square	f	Sig.
I	Regression	10.381	4	2.595	6.132	0.000
	Residual	57.985	137	0.423		
	Total	68.366	141			

Table 3.6 Tests of normality

	Kolmogorov–Smirnov[a]			Shapiro–Wilk		
	Statistic	df	Sig.	Statistic	df	Sig.
Unstandardized Residual	0.187	142	0.000	0.926	142	0.000

Table 3.7 Coefficients

Model		Unstandardized Coefficients		Standardized Coefficients	T	Sig.	Collinearity Statistics	
		B	Std. Error	Beta			Tolerance	Variance inflation
I	(Constant)	2.310	0.055		42.309	.000		
	Trading Tips	0.105	0.055	0.151	1.914	0.058	1.000	1.000
	Branch Facility	0.062	0.055	0.089	1.126	0.262	1.000	1.000
	Customer Handling	0.178	0.055	0.256	3.256	0.001	1.000	1.000
	Account Handling	0.164	0.055	0.236	2.999	0.003	1.000	1.000

3.6.4 Regression equation

As shown in Table 3.7, we construct the regression equation "ERR = (2.310) Constant + 0.105 (Trading Tips) + 0.062 (Branch Facility) + 0.178 (Customer Handling) + 0.164 (Account Handling)" where ERR is the expected rate of return.

3.7 ANALYSIS OF THE RESEARCH

The proposed study aims to investigate customer satisfaction towards the services provided by stock broking firms in Surat city. The analysis involves examining various factors such as customer awareness, perception, attitude, and satisfaction, and their associations with demographic variables and the expectation of return.

3.7.1 Demographic analysis

The data collected from 142 respondents indicates that the majority of respondents were male (97.2%) and primarily from the age group of 31 to 40 years (60%). Most respondents were graduates (90%) and engaged in business (35%) or salaried employment (28%).

3.7.2 Hypothesis 1: Association between awareness of future trading options and demographic variables

The chi-square test revealed significant associations between awareness of future trading options and demographic factors such as age, gender, income, and education. This implies that these demographic variables influence customers' awareness levels regarding future trading options.

3.7.3 Hypothesis 2: Factor analysis

The factor analysis was conducted to identify underlying factors affecting customer satisfaction. The analysis categorized the variables into four distinct groups: Trading Tips, Branch Facility, Customer Handling, and Account Handling. Each group represents a set of factors that contribute to customer satisfaction within the stock broking industry.

3.7.4 Hypothesis 3: Regression analysis

The regression analysis aimed to assess the impact of independent factors on customers' expectations of return. The results revealed that all independent factors, including Trading Tips, Branch Facility, Customer Handling, and Account Handling, significantly influenced customers' expectation of return from the equity market.

3.7.5 Conclusion in the context of Industry 5.0

In the evolving landscape of Industry 5.0, this research delves into the intricate realm of customer satisfaction within stock broking firms operating in Surat city. The analysis encompasses a multifaceted exploration of customer awareness, perception, attitude, and satisfaction, while also investigating their interplay with demographic variables and the anticipation of returns within the Industry 5.0 paradigm.

3.7.6 Demographic analysis

The comprehensive dataset, encompassing 142 respondents, unveils a demographic portrait dominated by male participants (97.2%) primarily within the age bracket of 31–40 years (60%). A substantial proportion of respondents are graduates (90%), engaged in either entrepreneurial ventures (35%) or salaried employment (28%).

3.7.7 Hypothesis 1: Association between awareness of future trading options and demographic variables

The scrutiny through chi-square examination reveals notable correlations between awareness of future trading options and pivotal demographic factors—age, gender, income, and education. Within the Industry 5.0 spectrum, this reinforces the nuanced influence of these demographic facets on fostering customer awareness regarding future trading alternatives.

3.7.8 Hypothesis 2: Factor analysis

Embedded within the tenets of Industry 5.0, factor analysis emerges as a pivotal tool, unravelling the intricate fabric of customer satisfaction dynamics. The segmentation into four distinct groups—Trading Tips, Branch Facility, Customer Handling, and Account Handling—offers a profound understanding of the constituents that collectively contribute to the customer satisfaction tapestry.

3.7.9 Hypothesis 3: Regression analysis

The Industry 5.0 landscape thrives on data-driven insights, and the regression analysis proves its mettle by illuminating the intricate relationships between independent factors and customers' expectations of return. Notably, within the Industry 5.0 ethos, factors such as Trading Tips, Branch Facility, Customer Handling, and Account Handling exhibit palpable and meaningful influence on customers' return anticipations.

3.8 CONCLUSION

Industry 5.0 serves as the backdrop for this research's conclusions, underscoring the transformational potential it brings to the stock broking domain. The interplay between customer awareness, satisfaction, and the futuristic essence of Industry 5.0 provides a rich tapestry of insights. The implications resonate across demographic dimensions, illuminating the intricate web of awareness creation. As Industry 5.0 steers the financial landscape into uncharted territories, the amalgamation of factor analysis and regression provides a compass for both service providers and market participants. The conclusions emanating from this study serve as guiding stars, steering the course for fostering customer satisfaction within the transformative currents of Industry 5.0.

3.9 FUTURE RECOMMENDATIONS

Dynamic customer engagement strategies: Given the evolving landscape of Industry 5.0, stock broking firms should adopt dynamic customer engagement strategies that harness emerging technologies. Personalized communication through AI-driven chatbots, interactive mobile apps, and augmented reality can enhance customer interactions and ensure a seamless trading experience.

Continuous customer education: As technological advancements reshape the stock broking landscape, there is a need for ongoing customer education. Firms should invest in comprehensive educational resources, webinars, and interactive tutorials to empower their investors with the knowledge and skills required to navigate Industry 5.0 effectively.

Predictive analytics for customer preferences: Leveraging data analytics and machine learning, stock broking firms can predict customer preferences and tailor their offerings accordingly. By understanding clients' trading habits, risk appetites, and investment goals, firms can provide more relevant and valuable services. The structure of organization and size of organization also play a crucial role (Vajpayee and Karthick 2019).

Ethical use of AI and automation: With the integration of AI and automation, ethical considerations become crucial. Firms should ensure transparent communication about the use of AI algorithms, automated trading, and data privacy. Establishing guidelines for responsible technology usage will build trust with customers as it has a strong link with human cognition and knowledge management (Vajpayee and Ramachandran 2019).

Enhanced security measures: As Industry 5.0 introduces advanced technologies, cyber threats also evolve. Stock broking firms must prioritize cybersecurity measures to protect customer data and financial transactions. Regular security audits and updates are essential to maintain a secure trading environment.

Holistic customer feedback mechanism: Firms must develop a holistic and real-time customer feedback mechanism that goes beyond traditional surveys. They should employ sentiment analysis tools to monitor social media, online forums, and customer interactions, gaining insights into customer sentiments and addressing concerns promptly for a positive organizational culture (Vajpayee and Chakraborty 2017; Vajpayee 2017; Chakraborty and Vajpayee 2017).

Collaborative ecosystem: Firms must foster partnerships with fintech startups, tech companies, and data analytic firms. Collaborative efforts can lead to innovative solutions, novel services, and a deeper understanding of customer needs within the industry 5.0 framework.

Investment in environmental, social, and governance (ESG) practices: With Industry 5.0's focus on sustainability and ethical practices, stock broking firms should align their operations with ESG principles. Incorporating ESG considerations into investment advice and offerings can attract socially responsible investors with all aspects of long-run happiness and relationship (Vajpayee and Sanghani 2022; Vajpayee et al. 2022).

Virtual reality trading platforms: Explore the potential of virtual reality (VR) and augmented reality (AR) to create immersive trading experiences. VR trading platforms can provide a three-dimensional view of market trends, enhancing decision-making processes for investors.

Regulatory adaptation: Industry 5.0 may lead to regulatory challenges as technology evolves. Stock broking firms should actively engage with regulators to ensure that new technologies are aligned with existing regulations or drive the creation of new regulatory frameworks.

Investor community platforms: Stock broking firms must develop online platforms where investors can share experiences, strategies, and insights (Chen and Lin 2014). Building a vibrant investor community can foster knowledge exchange, networking, and collaborative learning within the Industry 5.0 context.

AI-driven risk management: Firms must utilize AI algorithms to analyse market trends and anticipate potential risks. Advanced risk management tools can help investors make informed decisions, mitigate losses, and navigate the dynamic landscape of Industry 5.0.

In conclusion, the recommendations underscore the imperative for stock broking firms to embrace Industry 5.0's transformative potential (Bassey et al. 2011). By prioritizing customer engagement, technological innovation, ethical considerations, and collaborative partnerships, firms can position themselves as leaders in a rapidly evolving landscape while ensuring customer satisfaction and market efficiency.

REFERENCES

Agrawal, S. R., and D. Mittal. "Customer satisfaction, trust, and involvement in CRM implementation: A study in the stock broking industry.". Journal of Global Information Management (JGIM). 2019 Jan 1;27(1):144-64. www.indianjournals.com/ijor.aspx?target=ijor:ajrbem&volume=5&issue=5&article=001.

Bassey, N. E., U. E. Okon, and U. E. Umorok. "Effective customer service: A tool for client retention among stock broking firms in Nigeria." *African Journal of Business Management* 5, no. 20 (2011): 7987.

Chakraborty, D. K., and A. Vajpayee. "Impact of industrial relations practices on employee satisfaction and organization culture in manufacturing industries in Bhutan." *International Research Journal of Management Science and Technology* 8, no. 8 (2017): 57–65.

Chen, L., and H. Chen. "Personalized services and customer satisfaction in stock broking firms." *International Journal of Management Sciences* 32, no. 4 (2018): 78–95.

Chen, S. J., and B. Lin. "The mediating effect of e-satisfaction on e-service quality and e-loyalty." *Journal of Marketing Research* 47, no. 6 (2014): 1041–1056. doi:10.1509/jmkr.47.6.1041.

Gupta, R., and S. Agarwal. "Transparency in pricing and fee structures: Impact on customer satisfaction in the stock broking industry." *Journal of Consumer Behavior* 41, no. 1 (2017): 56–72.

Gupta, S., and R. Verma. "Impact of technological advancements on customer satisfaction in online trading platforms." *International Journal of E-Business Research* 16, no. 3 (2020): 45–62.

Jeyaprabha, D., and V. Sundar. "The impact of service quality on customer satisfaction in the stock broking industry.". *Revista Geintec-gestao Inovacao E Tecnologias, 11*(2), pp.931-940

Jeyaprabha B and Sunder C. What Influences Online Stock Traders' Online Loyalty Intention? The Moderating Role of Website Familiarity. *Journal of Tianjin University Science and Technology* ISSN (Online): 0493-2137 E-Publication: Vol: 55 Issue: 05:2022 Pp- 706-715.

Kajapriya, R., and S. Venkateswaran. Sequential Analysis of Indian Stock Broking Companies–A Survey. *International E- Research Journal.* 2021 Oct;20(30):27 Issue – 277 (B). Pp -64-67.

Lee, J., and M. Park. "Trust and customer satisfaction in the stock broking industry." *Journal of Business Ethics* 76, no. 2 (2019): 189–206.

Meenakshi, A. "Customer satisfaction towards motorcycles: A conceptual study. IMPACT: International Journal of Research in Humanities, Arts and Literature (IMPACT: IJRHAL) ISSN (P): 2347-4564; ISSN (E): 2321-8878 Vol. 2018;6:179-84.."

Rajkumar, P., and R. Ramaswamy. "Customer satisfaction in the stock broking industry: A review." *Journal of Marketing Research* 49, no. 7 (2017): 871–889. doi:10.1287/mnsc.49.7.871.16385.

Sharma, S., and N. Mehta. "Factors influencing Financial Backers' Exit Decisions from a New Venture. *Sustainability, Green Management, and Performance of SMEs.* 2023 Dec 4:33.."

Trivedi, S., Mehta, K. and Sharma, R., 2021. Systematic literature review on application of blockchain technology in E-finance and financial services. *Journal of technology management & innovation*, 16(3), pp.89–102.

Vajpayee, A. "A Comparative study of organizational culture in Indian multinationals and foreign multinationals of India." *International Journal of Indian Psychology* 4, no. 3 (2017): 112–122.

Vajpayee, A., and D. K. Chakraborty. "The societal culture of Bhutan and its impact on organizational culture, industrial relations, and employee satisfaction in manufacturing companies of Bhutan." *IOSR Journal of Business and Management* 19, no. 9 (2017) 1–7.

Vajpayee, A., and K. Karthick. "Organizational pyramid and size as a moderator variable in manufacturing industries of Bhutan." *International Journal of Innovative Technology and Exploring Engineering (IJITEE)* 8, no. 7S2 (May 2019) 503–509.

Vajpayee, A., and K. K. Ramachandran. *"Reconnoitring Artificial Intelligence in Knowledge Management." International Journal of Innovative Technology and Exploring Engineering (IJITEE)* 8, no. 7C (May 2019) ,114–117.

Vajpayee, A., and P. Sanghani. "Eternal happiness and endurance of life through Buddhism in Bhutan." *British Journal of Administrative Management* 52, no. 151 (2022): 10–23.

Vajpayee, A., P. Sanghani, D. Chakraborty, and A. Jain. "Doctrine of GNH and employee–employer relationship: A study of manufacturing industries of Bhutan." *Korea Review of International Studies* 15, no. 23 (2022): 23–38.

Wang, L., and Q. Li. "Customer engagement and satisfaction in the stock broking industry." *Journal of Marketing Analytics* 4, no. 2 (2016): 90–102.

Yang, Z., and X. Fang. "Online service quality dimensions and their relationships with satisfaction: A content analysis of customer reviews of securities brokerage services." *International Journal of Service Industry Management* 15, no. 3 (2004): 302–326.

Chapter 4

E-client safety as determinant of value in cyber-security aspect

Hasim Deari and Eldian Balla

4.1 INTRODUCTION

The PC has impacted the world in just about every avenue you can think of. Amazing developments in communication, collaboration, and efficiency. New kinds of entertainment and social media; access to information and the ability to give a voice to people who would never be heard of.

Bill Gates

As the world of business has changed its way of doing business in the electronic way on and the consumers' needs in turn, there is a crucial moment to estimate the value of service quality in the context of customer experiences buying from a corporate website. Customers have created a very impressive experience buying from internet and shaping their satisfaction from this process (Leong et al., 1998, Yang et al., 2003; Parasuraman and Zinkhan, 2002).

Zeithaml (2002) indicates that e-SQ has seven dimensions that form two scales: a core e-SQ scale and a recovery scale. Four dimensions – efficiency, reliability, fulfillment, and privacy – form the core e-SQ scale that can be used to measure customer perceptions of service quality. These dimensions and their definitions are: (1) Efficiency refers to the ability of the customers to get to the website, find their desired product and information associated with it, and check out with minimal effort. (2) Fulfillment incorporates accuracy of service promises, having products in stock, and delivering the products in the promised time. (3) Reliability is associated with the technical functioning of the site, particularly the extent to which it is available and functioning properly. (4) The privacy dimension includes assurance that shopping behavior data is not shared and that credit card information is secure.

DOI: 10.1201/9781032677040-4

4.2 FACTORS IMPACTING ON WEBSITE E-SQ

Zeithaml 2002) describes several criteria customers use when evaluating e-SQ and the quality of websites in general:

1. Information availability and content refers to the availability and depth of information; the ability to search price and quality information on the website; and the ability of the website's users to control the content, order, and duration (the amount of time the information is present) of product-relevant information.
2. Ease of use/usability refers to the idea that "customers' assessment of websites will likely be influenced by how easy the sites are to use and how effective they are in helping customers accomplish their tasks" and includes such website attributes as search functions, download speed, overall design, and organization.
3. Privacy and security are two distinguished criteria of a website and e-service quality. Privacy involves "the protection of personal information – not sharing personal information collected about consumers with other websites (as in selling lists), protecting anonymity, and providing informed consent" . Security involves "protecting users from the risk of fraud and financial loss from the use of their credit card or other financial information" , as well as the ability to provide data confidentiality, security auditing, encryption, and anti-virus protection.
4. Graphic style involves website attributes such as choice of colors, layout, print size and type, photographs, graphics, animation, three-dimensional (3D) effects, and multimedia.
5. Fulfillment/reliability is concerned with the actual performance of the company, rather than with the website performance; defined as the provider's ability to deliver the service or product as promised.
6. Access is the presence of the contact information on the company's website.
7. Responsiveness is the promptness with which the company's personnel give feedback to customers via e-mails.
8. Personalization refers to the website's ability to provide personalized and customized services according to customers' preferences.

4.3 WEBSITE CONVENIENCE

Companies in the last two decades, especially during the pandemic time, have changed their way of communication with consumers all around the world. They have tremendously increased the quality of websites for their consumers and offer a lot of information and security about different products and services. All these activities lead us to describe the concept of WEBQUAL (Table 4.1). According to Loino et al. (2002), this concept is

Table 4.1 The concept of WEBQUAL

Usefulness	Description
Informational Fit-to-Task	The information on the website is pretty much what I need to carry out my tasks.
	The website adequately meets my information needs.
	The information on the website is effective.
Interactivity	The website allows me to interact with it to receive tailored information.
	The website has interactive features, which help me accomplish my task.
	I can interact with the website to get information tailored to my specific needs.
Trust	I feel safe in my transactions with the website.
	I trust the website to keep my personal information safe.
	I trust the website administrators will not misuse my personal information.
Response Time	When I use the website there is very little waiting time between my actions and the website's response.
	The website loads quickly.
	The website takes long to load.

Easy of Use	**Description**
Ease of Understanding	The display pages within the website are easy to read.
	The text on the website is easy to read.
	The website labels are easy to understand.
Intuitive Operations	Learning to operate the website is easy for me.
	It would be easy for me to become skillful at using the website.
	I find the website easy to use.

Entertainment	**Description**
Visual Appeal	The website is visually pleasing.
	The website displays visually pleasing design.
	The website is visually appealing.
Innovativeness	The website is innovative.
	The website design is innovative.
	The website is creative.
Flow-Emotional Appeal	I feel happy when I use the website.
	I feel cheerful when I use the website.
	I feel sociable when I use the website.

Complimentary Relationship	
Consistent Image	The website projects an image consistent with the company's image.
	The website fits with my image of the company. The website's image matches that of the company.
On-Line Completeness	The website allows transactions on-line.
	All my business with the company can be completed via the website.
	Most all business processes can be completed via the website.

Source: Adapted from Loiacono et al. (2002, pp. 9–10).

made up of four components: usefulness, ease of use, entertainment, and complementary relationship, which include a range of website dimensions. To better understand this concept of WEBQUAL, we will describe all four dimensions by construct used by the same authors. The study conducted by Yang et al. (2005) has determined that there are five service quality dimensions perceived by users of an IP Web portal: usability, usefulness of content, adequacy of information, accessibility, and interaction. Each of the five identified and verified factors had a significant impact on overall service quality. Through understanding the service quality dimensions for IP Web portals, an organization will stand a much better chance of gaining more business and serving its stakeholders. Piccoli et al., (2004) stated that the use of IT allows organisations to create customer value through improved customer service. They argue that a firm's website represents *"the most visible instance of a Network-based Customer Service System (NCSS)" defined as "a network-based computerized information system that delivers service to a customer either directly (e.g., via a browser, PDA, or cell phone) or indirectly (e.g., via a service representative or agent accessing the system)"*

The dynamics of the last few years in the world of consumer buying have changed at a great pace, especially the period of COVID-19 almost changed a lot in the way of consumer buying. All this is the result of a change in the distribution network of the internet to a large extent, not only in developed countries but also in developing countries. Consumers life is better today than before due to the facilities and buying from home or feeling security during the process of buying online. According to Rita etal., (2019), customers can just sit at their homes, place their orders, pay via credit card, and wait until the goods are delivered to their homes. In addition, the internet has been generating consumer empowerment for over a decade (Pires et al., 2006).

4.4 THEORETICAL FRAMEWORK

4.4.1 What is e-commerce?

Searching by different sources about e-commerce there are many concepts that describe and explain the above terminology. The concept of e-commerce is not new for the developed countries and for their customers, but recently this concept is also not new for the rest of the world due to the changes of internet connection and internet networks.

There is one very important question about e-commerce, that is, what involves e-commerce? According to Hutt and Speh (2004), e-commerce involves "business communications transmissions over networks and through computers, specifically the buying and selling the goods, services, and the transfer of funds through digital communication." Customers can just sit at their homes, place their orders, pay via credit card, and wait until the goods are delivered to their homes (Rita et al., 2019).

Moreover, Tsao etal., (2016) studied the impact of e-service quality on online loyalty based on online shopping experience in Taiwan and showed that system quality and electronic service quality had significant effects on perceived value, which in turn had a significant influence on online loyalty.

A study conducted by Yang et al. (2005) explains that services provided by IP web portals generally consist of three types of interactions: (1) between customers and the portal employees via either Internet-based communication tools (e.g., email, chat room, etc.) or traditional channels (e.g., mail, fax, etc.); (2) between customers and the portal; and (3) among peer users of similar goods and services via email, chat rooms, etc. The quality issues concerning the first type of interactions are mainly traditional ones, while those of the second and third include web design. A study of the service quality of the portal needs to integrate both traditional and web design quality (the technical quality of a web site).

According to Kearney, (2015), retail e-commerce has grown nearly to US$ 840 billion in 2014 surpassing the sales of US$ 695 billion in year 2013 and it was estimated to increase to US$ 1506 billion in 2018. The continuous sales increment indicated that e-commerce has enormous market potential.

4.4.2 Type of e-commerce?

E-commerce is not just a theoretical concept. This concept takes on its importance by being classified into their types. These types of e-commerce explain in real time the importance of this business concept. Based on the Swedish Government Official Reports (1999:106), there are two types of e-commerce: indirect and direct.

Indirect e-commerce is when a consumer orders a product online, pays through an invoice or by cash on delivery, and gets the product delivered in a traditional way (Tables 4.2 and 4.3).

However, according to Misevičiūtė (2001), e-business can be supported by all market participants, such as:

- Business (B) with business (B): (B2B)
- Business (B) with government agencies (G): (B2G)
- Private and other business enterprises (B) with the budget, state-owned capital companies (G): (B2G)
- Public authorities and budget firms (G) with public institutions and budget firms: (G2G)
- Business (B), public authorities, and budget firms (G) with individual consumers and households (C), as well as by identifying buyers and recipients of electronic models: (B2C and G2C)
- Individual consumers and households with each other: (C2C).

Table 4.2 Electronic commerce definitions

No.	Authors	Definitions
1	Khan, A.G. (2016)	Electronic commerce, or e-commerce, is the buying and selling of goods and services on the internet. Other than buying and selling, many people use the internet as a source of information to compare prices or look at the latest products on offer before making a purchase online or at a traditional store.
2	Fichter, K. (2003)	E-commerce is understood as part of the business, which also includes, for example, video conferencing and teleworking. By definitions available so far, the term "e-business" can be defined as follows: business processes, commercial activities, or other economic tasks FORUM Fichter, Environmental Consequences of E-Commerce 27 conducted over the internet or computer-mediated networks (Intranet, etc.).
3	Shahriari, S., Shahriari, M., and Ggheiji, S. (2015)	Electronic commerce, commonly known as e-commerce, is trading in products or services using computer networks, such as the internet. Electronic commerce draws on technologies such as mobile commerce, electronic fund transfer, supply chain management, internet marketing, online transaction processing, electronic data interchange (EDI), inventory management systems, and automated data collection systems
4	Gunasekaran, A., Marri, H.B., McGaughey, R.E., and Nebhwani, M.D. (2002)	Electronic commerce provides new channels for the global marketing of tangible goods and presents opportunities to create new businesses providing information and other knowledge-based intangible products.

Source: Išoraitė and Miniotien (2018).

4.5 CHARACTERISTICS AND ADVANTAGES OF TRUST

Based on many academic studies, trust has received attention from scholars in the field of marketing, management, and even nowadays in psychology, economics, and other applied areas. The research on the trust concept comes from the analysis of personal relationship, in the field of social psychology, because it is considered an inherent characteristic of any valuable social interaction (Delgado and Munuera, 2001).

According to Doney and Cannon, (1997), marketing research on trust primarily focuses on two targets of trust: supplier firms and their salespeople. Trust of a supplier firm and trust of a supplier's salesperson, though related, represent different concepts. For example, a long-term relationship

Table 4.3 Advantages and disadvantages of buying online

Advantages	Disadvantages
Convenience – You save time by physically shopping. You can shop anywhere you have an internet connection and at any time.	**The ability to be cheated** – Your money security depends on your vigilance. So before submitting your personal information, check that the online store is trusted. When you buy online, you risk not getting or getting the same as the one shown in the photo.
Information details and recommendations – Detailed information is provided for each item in the e-shop. Here you can find product reviews and recommendations.	**It is easy to make a mistake on the price** – Consider that the product will have to pay more than specified.
Lower prices and better choices – Online sellers do not have to pay rent or wages to employees. It is convenient for you to compare prices for the same product in different e-shops in a short period.	**Lack of privacy** – Some online stores even send you unsolicited e-mails about various stocks and new items. On the one hand, this may seem useful, but on the other hand, it is a way to get you more money.
Starting an e-business requires less investment than traditional commerce.	When ordering a product online, it often takes a long time before it is delivered, and shipping charges sometimes exceed the price of the item, especially if the product is ordered from abroad. Repaying your online purchase may be more difficult than buying a traditional store.
More varied and more convenient supply of goods and services.	Often money will not be refunded for shipping, and in some cases, it will even have to cover the cost of returning the product itself. Also, when shopping online, it is important to make sure the website is trusted, and your credit card or online banking data will be protected. Finally, shopping online you will not be able to see the goods (like clothes).

Source: Išoraitė and Miniotienė (2018).

with a trusted supplier could be jeopardized by a company representative who proves to be dishonest and unreliable (e.g., Kelly and Schine, 1992).

According to Dwyer et., al. (1987), in channel settings, for instance, trust reduces tensions and conflicts between firms and facilitates information disclosure, thereby enhancing coordination and encouraging future transactions. In addition trust plays a critical role in customer–firm relationships as well, for it enhances and maintains consumer satisfaction and loyalty (Sirdeshmukh et., al. 2002). Trust is also vital in managing consumers' concerns about revealing personal information over the internet, the number one issue hampering the growth of e-commerce (Olivero and Lunt, 2004). Only by sustaining trust can marketers expect to establish enduring relationships with consumers, and it is by keeping a central focus on that idea that marketers build a value exchange that delivers consistent and progressive mutual benefits (Dayal et al.,2003).

Trust is a set of specific beliefs that reflect the confidence of consumers in an organization (McKnight and Chervany, 2002). Studying initial trust formation is important, because the results from such studies require an explanation beyond what calculative –based and knowledge based trust theories provide (McKnight et al.,1998).

According to Jeng Wu et al. (2010), in the context of a virtual community, trust is an important catalyst for facilitating social interaction and long-term relationships. According to Armstrong and Hagel (1997), virtual community members are expressly different from conventional e-shoppers because the social groups are able to balance the power associated with both vendors and customers through the release and exchange of information. When using new technologies, including the web and e-commerce, trust is considered to be important (Windham and Orton, 2000)

According to Atif (2002), to establish confidence between e-commerce parties there are three essential components as presented below:

- A simple model composed of a network of TSP – Trust Service Provider entities and an algorithm that establishes a trust path prior to carrying out an e-commerce transaction.
- An open extensible TSP architecture that can include additional e-commerce mediating services and an application programming interface (API) to run the trust-path building algorithm.
- A parallel implementation of the trust-path search algorithm, which contributes to a balance between time and network traffic complexity.

Dayal eta., (2003) in response to those security concerns, marketers are working to build trust with consumers through their online interactions. In addition, the authors state that the level of trust grows as marketers and consumers engage in a gradual "value exchange," through which consumers provide marketers with personal information and are rewarded in turn with products they actually want. The human perception of trust is a core ingredient in any online transaction (Andersen, 1998) and future e-commerce systems must support trust services to gain loyalty at both the consumer and provider ends (Manchala, 2000).

4.6 TRUST LEVEL IN THE TRANSACTION PHASE

A model of e-commerce customer relationship trust constructs is presented in Figure 4.1.

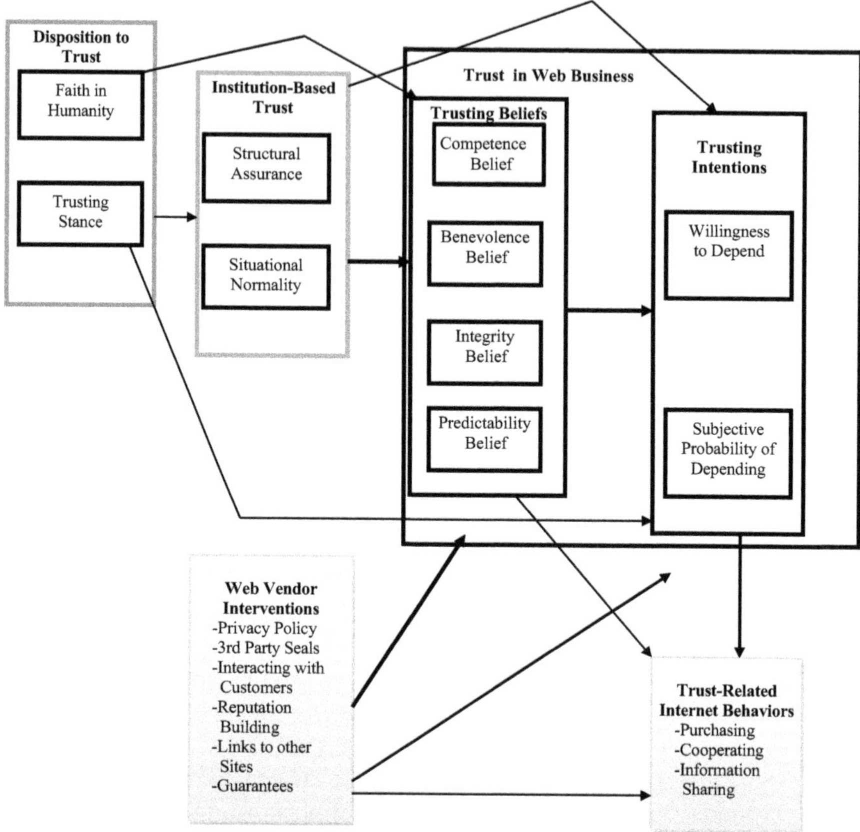

Figure 4.1 A model of e-commerce customer relationship trust constructs.

Note: Thinner arrows are proposed to be weaker links than thicker arrows, usually due to mediation effects.

Source: McKnight et al. (1998).

The model presented in Figure 4.1. clearly shows the main components that are important in building trust in a model of e-commerce customer relationship trust constructs. As we mentioned before, there are five dimensions of trust in this process and one web vendor. The first one is disposition of trust which incorporates two components, that is, faith in humanity and trust stance. In the second dimensions which belong institution-based trust, are incorporated two components that fulfill this dimension, i.e., structural assurance and situational normality. In the third one, there is one dimension which is named trust in web business where there are two main dimensions included in this part, i.e., dimension of trusting beliefs, benevolence belief, integrity belief, predictability belief, and trusting intentions, which take place willingness to depend and subjective probability of depending. The fourth one is trust related to internet behavior. In this dimension, there are components such as purchasing, cooperating, and information sharing.

4.7 INTERNET USAGE

Table 4.4 shows variability in the internet connection and penetration rate of the entire world North America has 93.4 % of penetration rate, with the second highest penetration rate in Europe (89.6 %); Latin America is in third place, 81.8 %; Middle East, 78.9 %; Oceania/Australia, 71.5 %; Asia, 67.4 %; and Africa 46.8 %. Based in the second variables included in table as a growth between 2000 and 2022: in the first place is Africa 14,362 %, in the second place is Middle East 6,378 %, third place is Latin America at 2,907 %, followed by Asia 2,467 %, Europe 614 %, Oceania/Australia 309 %, and North America 223 %.

The variable included in Table 4.4 is related to internet users. Worldwide, this variable is different and shows correlation with the number of populations in the world. Asia is the biggest internet user with 2.9 billion users, 750 million users, followed by Africa with 652 million, Latin America with 543 million users, North America with 349 million users, Middle East with 211 million users, and Oceania/Australia with 31 million users. Table 4.4 very well indicates that increasing internet usage worldwide leads to an increase in e-commerce and e-clients.

Table 4.4 World Internet usage and population statistics 2022 year estimates

World Regions	Population (2022 Est.)	Population % of World	Internet Users 30 June 2022	Penetration Rate (% Pop.)	Growth 2000–2022
Africa	1,394,588,547	17.6 %	652,865,628	46.8 %	14,362 %
Asia	4,352,169,960	54.9 %	2,934,186,678	67.4 %	2,467 %
Europe	837,472,045	10.6 %	750,045,495	89.6 %	614 %
Latin America/ Carib.	664,099,841	8.4 %	543,396,621	81.8 %	2,907 %
North America	374,226,482	4.7 %	349,572,583	93.4 %	223 %
Middle East	268,302,801	3.4 %	211,796,760	78.9 %	6,378 %
Oceania/ Australia	43,602,955	0.5 %	31,191,971	71.5 %	309 %
WORLD TOTAL	7,934,462,631	100.0 %	5,473,055,736	69.0 %	1,416 %

Source: www.internetworldstats.com/stats.htm (World Internet Usage and Population Statistics 2022 Year Estimates).

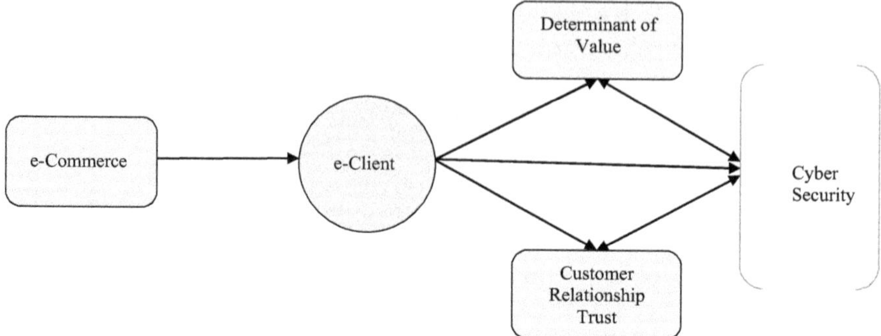

Figure 4.2 Authors' conceptual theoretical model.

4.8 CONCEPTUAL THEORETICAL MODEL

In this section, we present our research conceptual model (Figure 4.2). It's important to note that this model is designed by the authors of this research and is a unification of the theoretical concepts that we have described in this research.

This theoretical model represents the key components related to the topic of this chapter. In the conceptual model, we have incorporated the main features of each variable, respectively in the first line starting with e-commerce which is the first element that creates the entire conceptual model on one side, as well as a process of relationship between all elements presented above. Determinant of value and customer relationship trust are two very important dimensions that create a very powerful link between the e-customer and cyber security. This connection between these factors in the model creates a feedback for e-commerce companies.

4.9 CYBER SECURITY

More than ever in the last two decades, concern for cyber security has become one of the biggest priorities of many companies around the world, but not excluding the governments of different countries around the globe. The literature on cyber security is very rich in many aspects. According to Kaplan et al., (2011) and Bikoro et al., (2018), cyber security can be perceived as the digital protection of intellectual and commercial property against excessive use or unauthorized authorship. On the other hand, fighting against cybercrime is a delicate and costly mission (Bikoro et al., 2018). It is very important and disturbing on the other side that cyber-crime could cost the world 10.5 trillion dollars by 2025 (Morgan, 2020). This concern will increase the number of employees in the cyber-security job with

3.5 million jobs opening globally in 2025 (Morgan, 2022). Furthermore, Clinton (2022) states that calculating the precise economic impact of cyber-crime is extremely difficult because we are probably not aware of most cyber-attacks until well after they occur – if at all. It is extremely difficult to calculate the actual economic impact of stolen (usually digitally copied) intellectual property, and even when enterprises discover attacks they are often not reported publicly.

The more the concern about cyber security grows, the more countries of the world, but also multinational corporations, are organizing to fight and build defense strategies against cyber-attacks. Caravelli and Jones (2019) stated that, at the World Economic Forum, leaders in Davos, Switzerland, it was announced during a panel discussion that beginning in March 2018 in Geneva, a Global Center for Cyber Security would be established. In the press release, it was described as "the first global platform for governments, business, experts, and law enforcement to collaborate on cyber security." The Center will be supported by Interpol, the international law-enforcement entity. Alois Zwinggi, managing director of the World Economic Forum, will serve as the first director of the Center. Zwinggi said the Center's goals are:

- Consolidating existing cyber-security initiatives of the World Economic Forum
- Creating a library of cyber best practices
- Enhancing knowledge of cyber security
- Working toward an appropriate and agile regulatory framework on cyber-security practices
- Serving as a laboratory and early-warning think tank for future cyber-security scenarios.

Cyber security is a battle that is not just about control of hardware and software. It is every bit as much about the human factor behind the tools that make up cyberspace. In particular, there is an enormous body of evidence showing that human error can lead to successful hacking attacks and data breaches. The attendant costs, no matter how they are measured, are almost incalculable.

Caravelli and Jones (2019)

This confirms the fact that many cyber-attacks are increasingly an uncomfortable situation for individuals, hence the desire to take cyber-security measures (Wheeler, 2016, accessed 10 December 2022). According to Abbott (2019), the most effective cyber defense strategies are those that address and mitigate the vulnerabilities inherent in technology, business processes, and human behavior. He also states that organizations with advanced enterprise-wide cyber-security programs incorporate state-of-the-art digital security

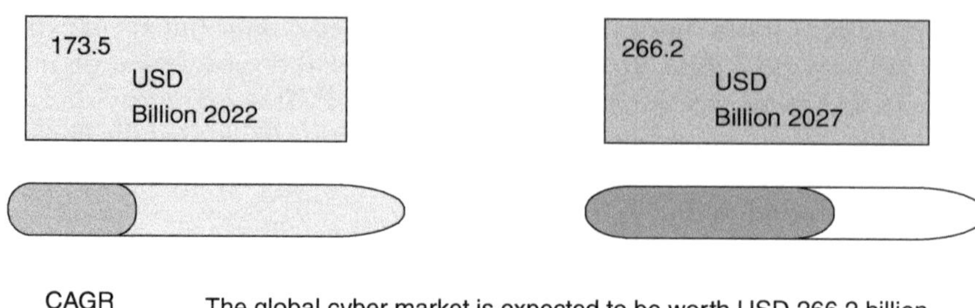

The global cyber market is expected to be worth USD 266.2 billion by 2027, growing at CARG of 8.9 % during the forecast period

Figure 4.3 Global cyber security market overview.

Source: MarketWatch, Accessed 10 December 2022.

Table 4.5 Companies in the global cyber-security market

Companies in the global cyber-security market	Name of company and country
	IBM (US), Cisco (US), Check Point (Israel), Trellix (US), Trend Micro (Japan), NortonLifeLock (US), Rapid7 (US), Micro Focus (UK), Microsoft (US), Amazon Web Services (US), Oracle (US), Fortinet (US), Palo Alto Networks (US), Accenture (Ireland), Cybe Ark (US), SentinelOne (US), Qualys (US), F-secure (Finland), F5 (US), DataVisor (US), RevBits (US), WiJungle (India), BluVector (US), Aristi Labs (India), Imperva (US), Securden (US), Forcepoint (US), Sophos (UK), RSA Security (US), Proofpoint (US), Juniper Networks (US), Splunk (US), SonicWall US), AlgoSec (US), Zscaler (US), Cynet (Israel), and Nazomi Networks (US)

Source: MarketWatch (Accessed 10 December 2022).

tools with effective human resource management techniques to optimize a security-minded culture and mitigate the vulnerabilities created by human errors (Figure 4.3).

Based on the MarketWatch report there are a number of companies worldwide that are involved in the cyber-security market. Table 4.5 presents a number of companies in the global cyber-security market.

4.10 RESEARCH METHODOLOGY

4.10.1 Qualitative and quantitative methods

In this research, we have used mixed methods, that is, qualitative and quantitative research (Jick, 1979; Venkatesh et al., 2013). There are essential distinctions between qualitative and quantitative methods. If the data is coded with numbers and the numbers are analyzed with statistical methods, this procedure is described as quantitative. On the other hand, if the data is described by text, diagrams, and the text and diagrams are interpreted, this level of research is described as the qualitative approach (Axinn and Pearce, 2006).

Patton (2002) states that the advantage of a quantitative method consists of its possibility to measure the reactions of a great many people to a limited set of questions, thus facilitating comparison and statistical aggregation of the data.

Qualitative data is the most important part of every research field. According to Patton (2002), there are three kinds of qualitative data:

- Interviews – refers to open-ended questions and probes that yield in-depth responses about people's experiences, perceptions, opinions, feelings, and knowledge.
- Observation – refers to fieldwork descriptions of activities, behaviors, actions, conversations, interpersonal interactions, organizational processes, or any other aspect of observable human experience.
- Documents – refer to written materials and other documents from organizational, clinical, or program records, official publications, and written responses to open-ended surveys or photographs.

4.10.2 Questionnaire design and data collections

In this research study, we have used the quantitative survey method by distributing the questionnaires to both undergraduate and postgraduate students in higher institutions and different participants.

In addition, close-ended questions with a 5-point Likert-type scale were used throughout the study; meanwhile, we have incorporated demographic variables to get more reliable data from the consumer perspectives. A multi-stage sampling method which combined both stratified and systematic sampling techniques was applied in this study. We have tested the reliability and validity of the study design, as well as 395 questionnaires were distributed to the respondents, and 350 questionnaires were collected, indicating an 88.6 % response rate. The data was analyzed and interpreted using SPSS software.

The collected data can be divided into two different types: primary and secondary data. Primary data is information that the researcher himself has collected through, for example, interviews and observations. Secondary data is material that has been collected earlier by someone else other than the researcher. In this case, the data comes from, for example, literature, articles, and different types of registers (Andersen, 1998).

4.11 EMPIRICAL ANALYSIS AND HYPOTHESES TESTING

The six hypotheses for this study were tested using t-test, that is, independent simple test based on the formula given below:

$$t = \frac{\overline{X}_1 - \overline{X}_2}{\sqrt{\left(\dfrac{(N_1-1)s_1^2 + (N_2-1)s_2^2}{N_1 + N_2 - 2}\right)\left(\dfrac{1}{N_1} + \dfrac{1}{N_2}\right)}} \qquad [1]$$

Where:

Degrees of freedom is 2 $N_1 + N_2 - X_1$ and X_2 are the respective sample means of the two groups

S_1 and S_2 are the standard deviations

N_1 and N_2 are the sample sizes of the two groups.

Figure 4.4 presents the authors' model of testing hypothesis.

- H: 1 E-clients believe in the online transactions.
- H: 2 E-clients are under pressure when they do electronic transactions.
- H: 3 E-clients trust cyber security when they do any electronic transaction.
- H: 4 E-clients consider the security of payments as a factor that impacts e-client safety as a determinant of value in CSA – Cyber security aspect
- H: 5 E-clients consider lack of personal and social contacts impact e-client safety as determinant of value in CSA
- H: 6 E-clients consider lack of knowledge and experience impact e-client safety as determinant of value in CSA

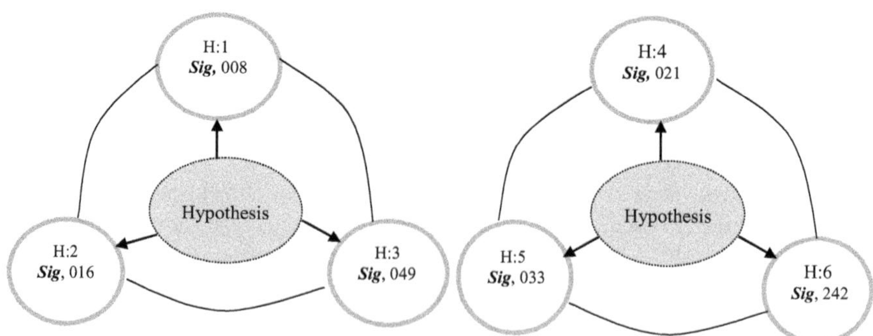

Figure 4.4 Authors' model of testing hypothesis.

4.12 CONCLUSIONS

Cyber security as we can see in many studies and research in different fields is the challenge for the twenty-first century for all communities worldwide, particularly for all these companies and organizations that do their activities electronically.

In particular, this study gives a better understanding of e-clients in the process of buying online and how they feel in this process of buying. This study makes it possible to lay a theoretical theory for the relationship between e-clients, trust, buying online, and cyber security.

Based on the results that we got from the primary data we can conclude that consumers are really concerned when they do any online transaction due to cyber-attacks, e-clients strongly believe in online transactions, e-clients trust the cyber security, e-clients consider the security of payments is a factor that impacts e-client safety and e-clients consider lack of personal and social contacts impact e-client safety as a determinant of value in CSA.

We can conclude that all five hypotheses have a level of significance less than 0.05, respectively one hypothesis has a bigger level of significance than 0.05, > (*Sig*, 242), i.e., e-clients consider lack of knowledge and experience impact e-client safety as a determinant of value in CSA.

These results also validate the existing findings of the related study. Most of the past findings agreed that there was no significant difference between males and females in terms of trust in online transactions. Also, these results recommend to business communities that consumer interest in buying online shows that e-clients have a continuous interest in online shopping, but, on the other hand, they are constantly concerned about their safety. Therefore, companies must protect consumers from continuous cyber-attacks.

REFERENCES

Abbott, J. (2019). Five ways HR can improve cyber security [blog]. Personnel Today. www.personneltoday.com/hr/hr-role-in-cyber-security

Andersen, I. (1998). Den uppenbara verkligheten – Val av samhällsvetenskaplig metod. Studentlitteratur, Lund, Sweden.

Armstrong, A. and Hagel, J. (1997). Net Gain: Expanding Markets Through Virtual Communities. Harvard Business School Press, USA.

Atif, Y (2002) Building trust in e-commerce. IEEE Internet Computing. Vol. 6, No.1, pp.18–24.

Axinn, W. and Pearce, L. (2006). Mixed Method Data Collection Strategies. Cambridge University Press, USA.

Ballester, E.D. and Munuera-Alemán, J.-L. (2001). Brand trust in the context of consumer loyalty. *European Journal of Marketing*, 35(11/12), 1238–1258.

Bikoro, A.D.M., Wamba, S.F ., and Robert, J.K.K. (2018). Determinants of cyber security use and behavioral intention: Case of the Cameroonian public administration. In: Rocha, Á., Adeli, H., Reis, L., Costanzo, S. (eds) Trends and

Advances in Information Systems and Technologies. WorldCIST'18 2018. Advances in Intelligent Systems and Computing, vol 746. Springer. https://doi.org/10.1007/978-3-319-77712-2_104

Caravelli, J. and Jones, N. (2019) Cyber Security Threats and Responses for Government and Business. Praeger, USA.

Clinton, L. (2022). Cybersecurity for Business Organization – Wide Strategies to Ensure Cyber Risk Is Not Just an IT Issue. Kogan Page Limited.

Dayal, S., Landesberg, H., Zeisser, M. (2003). How to Build Trust Online. In: Petrovic, O., Ksela, M., Fallenböck, M., Kittl, C. (eds) Trust in the Network Economy. Evolaris, vol 2. Springer, Vienna. https://doi.org/10.1007/978-3-7091-6088-6_5

Doney, P.M. and Cannon, J.P. (1997) An examination of the nature of trust in buyer–seller relationships. *Journal of Marketing*, 61(2), 35–51.

Dwyer, F.R., Paul, H.S., and Sejo, O. (1987). Developing buyer–seller relationships. *Journal of Marketing*, 51(2), 11–28.

Fichter, K. (2003). E-commerce sorting out the environmental consequences. *Journal of Industrial Ecology*, 6(2).

Gunasekaran, A., Marri, H.B., McGaughey, R.E., and Nebhwani, M.D. (2002). E-commerce and its impact on operations management. *International Journal of Production Economics*,75, 185–197.

Hutt, M. Speh, Th. (2004) Business Marketing Management: A Strategic View of Industrial and Organizational Markets. Thomson/South-Western,USA.

Išoraitė, M. and Miniotien, N. (2018). Electronic commerce: Theory and practice. *Integrated Journal of Business and Economics*, 2(2), 73–79.

Jick, T.D. (1979) Mixing qualitative and quantitative methods: Triangulation in action. *Administrative Science Quarterly* 24(4), 602–611.

Kaplan, J., Sharma, S., and Weinberg, A. (2011). Meeting the Cybersecurity Challenge. www.mckinsey.com/capabilities/mckinsey-digital/our-insights/meeting-the-cybersecurity-challenge#/)

Kearney, A.T. (2015). The 2015 Global Retail E-Commerce Index: Global Retail E-Commerce Keeps on Clicking. A.T. Kearney, Chicago.

Kelly, K. and Schine, E. (1992). How did sears blow this gasket? *Business Week* (June 29), 38.

Khan, A.G. (2016). Electronic commerce: A study on benefits and challenges in an emerging economy. *Global Journal of Management and Business Research: B Economics and Commerce*, 16(1).

Leong, E.K.F., Huang, X., and Stanners, P.-J. (1998). Comparing the effectiveness of the web site with traditional media. *Journal of Advertising Research*, Sept.–Oct., 44–51.

Loiacono, E., Watson, R.T., and Goodhue, D.L (2002). WEBQUAL: A measure of website quality. In AMA Winter Conference, Austin, Texas, USA.

Manchala, D. (2000). E-commerce trust metrics and models. *IEEE Internet Computing*, 4(2), 36–44.

Margarita, I. and Neringa, M. (2018). Electronic commerce: Theory and practice. *Integrated Journal of Business and Economics*, 2(2), 73–79.

McKnight, D.H. and Chervany, N.L. (2002). What trust means in e-commerce consumer relationships: An interdisciplinary conceptual typology. *International Journal of Electronic Commerce*, 6(2), 35–59.

McKnight, D.H., Cummings, L.L., and Chervany, N.L. (1998). Initial trust formation in new organizational relationships. *Academy of Management Review*, 23, 473–490.

Misevičiūtė, B. (2001). Elektroninė komercija [PPT]. Retrieved 2018, from http:// kopustas.elen.ktu.lt/studentai/lib/exe/fetch.php?media=elektronine_komerc ija.ppt

Olivero, N. and Lunt, P. (2004). Privacy versus willingness to disclose in e-commerce exchanges: The effect of risk awareness on the relative role of trust and control. *Journal of Economic Psychology*, 25(2), 243–262.

Parasuraman, A. and Zinkhan, G.M. (2002). Marketing to and serving customer through the internet: An overview and research agenda. *Journal of the Academy of Marketing Science*, 30(4), 286–295.

Patton, M. (2002). Qualitative Research & Evaluation Methods, 3rd edn., Sage Publication, USA.

Piccoli, G., Broham, M.K., Watson, R.T., and Parasuraman, A. (2004). Net-based customer service systems: Evolution and revolution in web site functionalities. *Decision Sciences*, 35(3), 423–455.

Pires, G. South, P. and Rita, P. (2006) The internet, consumer empowerment and marketing strategies. *European Journal of Marketing*, Vol. 40 No. 9/10, pp. 936–949.

Rita, P., Oliveira, T., and Farisa, A. (2019). The impact of e-service quality and customer satisfaction on customer behavior in online shopping. *Heliyon 5*, e02690. https://doi.org/10.1016/j.heliyon.2019.e02690.

Shahriari, S., Shahriari, M., and Ggheiji, S. (2015). E-commerce and it impacts on global trend and market. *International Journal of Research-Granthaalayah*, 3(4), 49–55.

Sirdeshmukh, D., Jagdip, S., and Sabol, B. (2002). Consumer trust, value, and loyalty in relational exchanges. *Journal of Marketing*, 66(1), 15–37.

Swedish Government Official Reports (1999:106). Consumer and IT an Inquiry Into Data, Trade and Marketing. Norsted's Printing House, Stochholm.

Tsao, W.-C., Hsieh, M.-T., and Lin, T.M.Y. (2016). Intensifying online loyalty! The power of website quality and the perceived value of consumer/seller relationship. *Industrial Management & Data Systems*, 116(9), 1987–2010.

Venkatesh, V., Brown, S.A., and Bala, H. (2013). Bridging the qualitative-quantitative divide: Guidelines for conducting mixed methods research in information systems. *MIS Quarterly*, 37(1), 21–54.

Windham, L. Orton, K. (2000) The Soul of the New Consumer: The Attitudes, Behavior, and Preferences of E-Customers. New York. Allworth Press.

Yang, Z., Cai, S., Zhou, Z., and Zhou, N. (2005). Development and validation of an instrument to measure user perceived service quality of information presenting web portals. *Information and Management*, 42, 575–589.

Zeithaml, V. (2002). Service excellence in electronic channels. *Managing Service Quality*, 12(13),135–138.

Internet sources

www.internetworldstats.com/stats.htm
https://cybersecurityventures.com/hackerpocalypse-cybercrime-report-2016/
www.marketsandmarkets.com/Market-Reports/cyber-security-market
    505.html?gclid=CjwKCAiA-dCcBhBQEiwAeWidtaTD1E2GCpay-
    aVeuoCH9AAZ4K44JFTkNlSGc7-IWJzXmGiGAo8DIRoCsokQAvD_BwE
    (Accessed10 December 2022).
https://cybersecurityventures.com/cybercrime-damage-costs-10-trillion-by-2025/#
https://blogs.gartner.com/john-wheeler/gartner-top-ten-cybersecurity-predicts/
    (Accessed 10 December 2022).

Chapter 5

Overview of Industry 5.0 and its impact on HR

A comprehensive review

Vidhi Tyagi, Shikha Mittal, and Abha Gupta

5.1 INTRODUCTION: BACKGROUND AND RATIONALE

Industry 5.0 puts the welfare of the workforce at the center of the whole process of an organization to overcome the human resource difficulties of Industry 4.0. After two virtual workshops, where ideas were discussed and shared, the concept of Industry 5.0 evolved in 2020. It was formally published in a document by the European Commission (EC) in January 2021. The goal of the EC paper is to promote transformation and changes in organizations and sectors so that organizations become more human-centric and sustainable. Industry 5.0 is aligned with European societal objectives, i.e., industrial sustainability must be ensured by respecting the limits of our planet and the welfare of industrial workers in addition to job creation and resilient development. Resilience, sustainability, and human centricity are the three main pillars of the Industry 5.0 approach, which is focused on people (Nahavandi, Saeid, 2019).

Sindhwani et al (2022) the goal of Industry 5.0 is to fully use cutting-edge digital technologies and the human–machine interface making the environment of the organization more manpower friendly. In today's smart industrial environments, the operator works alongside and with the aid of equipment. With the implementation of Industry 5.0, the EC hopes to increase workplace inclusivity while also fostering more resilient and sustainable working practices. The topic is covered in stages: the Industry 5.0 operator; the evolving role of HR experts and the necessity for strategic adaptation in this digital age; and, finally, specific challenges, constraints, and the future research agenda are brought up. Therefore, it is possible to evaluate Industry 5.0's impact on HR practices and decide whether it is truly human-oriented.

DOI: 10.1201/9781032677040-5

5.1.1 Objectives of the study

- To identify the changing role of HR professionals in an ever-evolving Industry 5.0
- To explore specific areas of HR that are affected by Industry 5.0
- To explore the changing workforce and competencies required for HR in Industry 5.0

5.1.2 Scope and methodology

Organizational performance, which ultimately results in organizational success, is greatly influenced by decisions made by the human resources (HR) department. The organizational structure, the levels of workers, and their significance in carrying out organizational tasks are all influenced by how HR practices are managed (Betchoo K. Nirmal, 2016)Organizations looking to improve the efficacy and efficiency of their human resource management (HRM) activities by utilizing the Industry 5.0 strategy will find the study to be relevant and helpful. To know the impact of HR on Industry 5.0, a secondary research approach was adopted.

5.2 UNDERSTANDING INDUSTRY 5.0

5.2.1 Definition and evolution of industrial revolutions

As Industry 5.0 is still in the evolution phase, a large number of definitions have been coined by industry experts and various researchers. According to Michael Rada, the organization's founder and leader, Industry 5.0 is the first industrial evolution to be driven by humans and is based on the 6R (Recognize, Reconsider, Realize, Reduce, Reuse, and Recycle) principles of industrial upcycling, a methodology based on better waste management and efficient design of logistics (Yevsieiev and Gurin, 2023)

According to S. Nahavandi (2019) and Demir et al (2019) , Industry 5.0 reaffirms the use of workers in the workplace, where humans and machines work together to improve process efficiency by utilizing the ingenuity and brainpower of human labor. According to Aslam et al (2020) and Friedman and Hendry (2019) , Industry 5.0 forces business professionals, information technologists, and philosophers to concentrate on the integration of human elements into industrial system technology.

According to the criteria provided by the researchers, Industry 5.0 is the next industrial revolution, where humans and robots co-work, not an evolution (Schwab 2017) Matching human intellect with machine intelligence and teaching cobots to adapt to a radical transformation of the human brain while co-working are the key implications (Maddikunta, Praveen Kumar Reddy, Quoc-Viet Pham, B. Prabadevi, Natarajan Deepa,

Kapal Dev, Thippa Reddy Gadekallu, Rukhsana Ruby, and Madhusanka Liyanage,2022). Therefore, cobots can be used for labor-intensive tasks and human intelligence can be used to critically evaluate the reasoning behind customization, relieving the need for a labor force that is already underutilized.

Below is a brief description of the industrial revolution-

INDUSTRY 1.0: This was the most notable revolution in the manufacturing sector with a significant increase in productivity, also was the beginning of the first Industrial era. With the evolution of steam power and engines based on steam around the 18th century, production of thread on single spinning wheels was increased up to 8 times against the manual spinning wheels.

INDUSTRY 2.0: Next phase of the industrial revolution occurred between 1871 and 1914, known as Industry 2.0. This industrial revolution was driven by innovative ideas, involving more advanced transfer of persons and better machines for manufacturing. This led to high economic growth and increased productivity, but it also caused unemployment due to increased use of machines for better productivity.

INDUSTRY 3.0: Industrial Revolution 3.0, started with the advent of digitalization era. It started in the late 70s in the 20th century, with the increased use of automation using program-based controlled computers. It helped in mass production of goods, using digital logics and integrated circuit chips. It was driven by the increased use of computers, cellular phones, and internet.

INDUSTRY 4.0: Industrial Revolution 4.0 was driven by the combination of better machines and advanced technology like artificial intelligence (AI), Internet of Things (IoT), robots, cloud computing, three-dimensional (3D) printing, etc. With the advancement of technology, the industry has become more flexible and driven based on data.

With the evolution of technology, efficient and intelligent machines are the drivers of Industrial Revolution 5.0.

5.2.2 Core features and principles of Industry 5.0

Industry 5.0 (IR 5.0) is a new manufacturing age, marked by the fusion of cutting-edge technological innovation with human ingenuity to provide a more cooperative and adaptable industrial environment. Human-centricity, sustainability, and resilience are Industry 5.0's three main tenets.

Human-Centricity: In IR 5.0, robot speed, efficiency, and consistency are blended with human innovation and artistry. As a result, it supports variety, talent, and human empowerment.

Sustainability: One of the most prominent aspects of IR 5.0 is additive manufacturing, also referred to as 3D printing, which is utilized to increase the sustainability of product manufacturing. With IR 5.0, incentives were added to products and services to increase consumer satisfaction.

Resilience: The term "resilience" relates to the requirement to increase the durability of industrial output. When humans and robots work together, high resilience can be achieved.

Promoting employee happiness and job satisfaction is an essential component of Industry 5.0's human factor. Industry 5.0 aims to empower workers and improve their well-being by fostering a more supportive and positive work environment with technology like digital assistants and modern safety systems. As a result, employees may feel more involved and valued in their work, which can enhance job satisfaction, productivity, and innovation. Technologies like virtual reality (VR) and augmented reality (AR) make visualization encouraging and testing of a new product very easy. It also gives confidence to workers to be more innovative and creative.

Machines and people are viewed as complementary rather than as distinct entities in Industry 5.0. Advanced technologies like VR, AR, real-time data flow, and robotics make it easy to express ideas and design products out of their imagination, while machines take care of the monotonous and tedious jobs. This cooperation results in a manufacturing process that is more responsive and adaptable to shifting market demands and adaptations.

5.2.3 Key technologies enabling Industry 5.0

By adopting technical trends like EC, Digital Twins (DT), Internet of Everything (IOE), big data analytics, cobots, 6G, block-chain, Network Slicing (NS), Extended Reality (XR), and Private Mobile Network (PMN), Industry 5.0 plays a key role in a vast range of applications [5]. All these enabling technologies combined with innovation and logical abilities are used to boost productivity and provide customized goods more swiftly by the organizations.

- Industry 5.0 reduces any delay in the process, reduces the bandwidth requirement, and helps enhance the security and privacy of data. Edge computing has made transactions possible with limited connectivity. In Industrial Revolution 5.0, edge computing enables standard hardware and software to share and process the data seamlessly about the relative sector. The DT enables Industry 5.0 to resolve technical issues by recognizing issues more quickly, locating the scope of upgradation or reconfiguration based on productivity, producing more precise projections, anticipating faults in the future, and preventing significant financial losses.

- In Industry 5.0, cobots use robots in collaboration with humans, which can be used for increased productivity. It makes humans and robots more interactive and safe to work in close proximity, allowing for more interesting responsibilities for human workers.
- The Internet of Things (IoT) involvement in Industry 5.0 offers the chance to lower operational costs by removing communication channel bottlenecks, lowering latency, cutting supply chain waste, and improving production procedures.
- Industry 5.0 operations that require mass customization benefit from big data analytics' zero-fail integration with the available resources. Manufacturers can generate and manage massive volumes of data with the help of real-time analytical data shared with intelligent systems and data centers.
- Smart contracts and blockchains are being used to make contracting more dynamic and to automate the procedures through which different stakeholders come to agreements. For Industry 5.0 applications, smart contracts are employed for security enforcement, including authentication and automated service-oriented operations.
- 6G networks will allow transfer of data at ultra-high speed, ultra-low delay in data transfer, ultra-high reliability, will be highly energy efficient, will have high traffic capacity, etc. for Industry 5.0 applications, in line with the standards of the intelligent information society.
- NS offers several virtualized networks for IoT networks' network monitoring at a cheap cost and with efficient network resource utilization. Industry 5.0 uses for XR include remote help, real-time assembly line monitoring, healthcare education, remote healthcare, navigation, training of driving and flying planes, to give product experience in retail, etc. To provide location-specific connection solutions for Industry 5.0 applications including factories, hospitals, schools, and universities, PMN are utilized to supply localized, case-specific network services.

5.3 INDUSTRY 5.0 AND ITS IMPLICATIONS FOR HR: THE EMERGENCE OF INDUSTRY 5.0 AND HR TRANSFORMATION

One of the key responsibilities of HR departments is to carry out work analyses, create job descriptions, and fill open positions with qualified candidates. Except for those sectors that use industrial robots, the great majority of businesses now employ people with the growth of Industry 5.0. HR departments will encounter new difficulties as robots become an integral component of organizations. They will need to determine the jobs that robots will take over in addition to their current duties. In essence, they will choose which jobs to assign to robots. HR will become more significant here in defining the jobs. The HR department will eventually change, and might even have new names in the future.(DiRomaualdo et al. 2018)

5.3.1 Shifting HR practices and strategies in the digital age

Digitalization of HR practices is accelerating, which is changing how it manages daily operations. The change encourages the reduction of manual interventions and other time-consuming paperwork processes, while also boosting operational resilience and productivity. The employee now occupies a new position inside the business because of Industry 5.0, and the owners view them as an investment. This tactic enables both the employer and the employee to advance because the recruiting organization is interested in investing in the abilities, competencies, and well-being of the employees to realize shared goals.

5.3.2 Digital HRM processes

With the advent of sophisticated technology and "Digital operations thinking," organizations need to rethink the HRM operations key responsibility areas (KRAs) and need to develop new HRM processes which will define the whole framework end-to-end for better results. The effects of these changes will be visible at every step starting from selection to development and training, to better performance standards and service quality. (Fenech, Roberta, 2022)

5.3.3 Talent selection

The key area which has widely changed with digital technology is recruitment and staffing in HRM organizations. Digital HRM has made the process of selecting candidates easier, and it helps in defining the required skill set for vacant positions and shortlisting the candidates based on that.

With internet becoming more and more useful in HRM processes, organizations have also reinvented the whole selection process; these days it is a traditional process such as job posting in newspapers. These days professional networking websites and employment websites have become better options due to increased use of internet.

5.3.4 Training and development

With the internet becoming widely used, personalized training became common in the late 1990s and early 2000s, and the momentum of digital education became very popular. HR department has changed widely due to technological advancement globally. Advanced digital technologies help in seamless integration and high flexibility. Training and development structure has totally changed with the advent of digital technology. With the new technology, digital training has developed a new system as per the requirement of the organization. It has become possible to create a pool of trained manpower with a varied skill set.

5.3.5 Evaluative functions

Assessment and motivation are the basic pillars of the evaluation function in HRM. With advanced technology and digitization, evaluation has become more accurate, as data is left on data platforms. This process is becoming totally data-driven and data is also being recorded on data platforms. The whole process at each and every step is data-driven, all the decisions are taken based on data. All the data related to an employee is recorded at each step and converted to digital form for integration of all the data in the system. The database has all the data on employee performance, turnover, and absenteeism, aligned with organizational goals. Developing such a database can be of significant importance for HRM. Capturing and integrating the data in a database and taking the trends into consideration based on a comparison of data with old and current trends can help in policymaking for HRM.

5.3.6 Managing remote teams

The biggest change driven by digital technology is the management of remote workplaces. With work from home becoming a new normal in various industries, a few sectors have decided to shift to a permanent remote working culture. These changes in the process have posed new challenges for HR and require big changes in the HR structure. HR professionals have been assigned a big task of employee management from a remote workplace, which lacks face-to-face interaction with the workforce.

5.3.7 Automation and resource management

Digital technology has helped in increasing the use of automation in HR processes, has created more resources, and made resources management better. All the data flow is now digitized, and various processes have been automated like onboarding of employees can be automated rather than sending papers for acceptance by employees.

5.4 IMPACT ON TALENT ACQUISITION AND MANAGEMENT: TALENT ACQUISITION STRATEGIES TO INDUSTRY 5.0

In the wake of Industry 5.0, the landscape of talent acquisition and management has undergone a profound transformation. Such transformations have made an impact on the skills and competencies required in the industry which calls for upskilling and reskilling the workforce and adapting talent acquisition strategies to the demands of the industry.

Recruitment applications, software, and chatbots are used for preliminary screening of candidates to ease the process, reduce manual effort, and speed up the process. Organizations are implementing hybrid recruitment models that combine traditional methods with cutting-edge technology to ensure hiring of the best talent.

Big data analytics plays a significant role in identifying and assessing potential candidates. Further, predictive analytics is used to make informed decisions about which candidates are the best fit for a role. VR and AR are being used to enhance the candidate experience and participate in

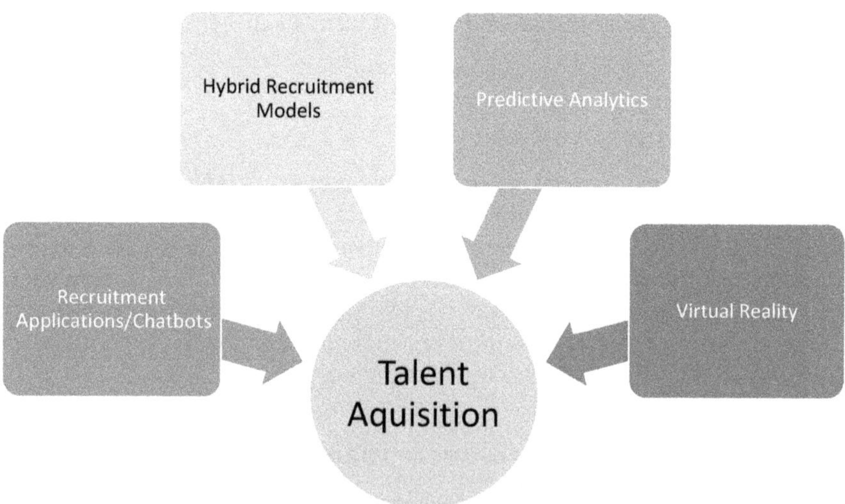

Figure 5.1 Talent acquisition strategies to Industry 5.0.

Source: Author.

realistic job previews. Organizations even conduct interviews in virtual environments and are now looking for candidates who can prosper in a virtual environment. Industry 5.0 has enhanced the adoption of remote work. Figure 5.1 highlights some of the talent acquisition strategies for Industry 5.0.

5.4.1 Skills and competencies in demand in Industry 5.0

In the light of Industry 5.0 where humans and machines work together, working with data and understanding data analytics has become crucial. Data-driven decision-making has been in trend in almost every function of management. Organizations demand for employees who can work with data and understand data analytics. Employees are expected to be comfortable with data-driven decision-making and proficient in data analysis tools. AI and machine learning concepts and their applications have become valuable across industries, as these technologies have become integral to various processes.

With the rapid change in technology and adoption of new working techniques, the ability to adapt and learn new skills has become a fundamental competency. Recruiters focus on digital literacy and digital proficiency while employing talent along and focusing on their capability to adapt to new technologies. Further, the problem-solving and critical thinking skills are in demand as automation handles routine tasks. Organizations look for candidates who are competent in analyzing complex problems and devising effective solutions as a core competency, especially when automation handles routine tasks.(Grant, Adam M., Yitzhak Fried, and Tina Juillerat, 2011).

Organizations cannot deny the human aspect and so soft skills like communication, empathy, and teamwork remain critical as humans continue to collaborate with machines. Industry 5.0 requires employees to work across different disciplines, interdisciplinary knowledge, and the ability to collaborate with experts is also relevant. Organizations look for interdisciplinary skills and collaboration and demand for people who can bring in diverse perspectives to the table and work in collaboration. Figure 5.2 illustrates skills and competencies demanded by Industry 5.0.

Industry 5.0 places a premium on creativity and innovation. Individuals who can think creatively and contribute new ideas are valuable assets. A strong sense of ethics and social responsibility is crucial for organizations and their HR departments to ensure the responsible use of advanced technologies (Amabile M. Teresa 2019) The need for cybersecurity expertise also grows with increasing digitalization. Professionals who can protect an organization's digital assets are highly sought after.

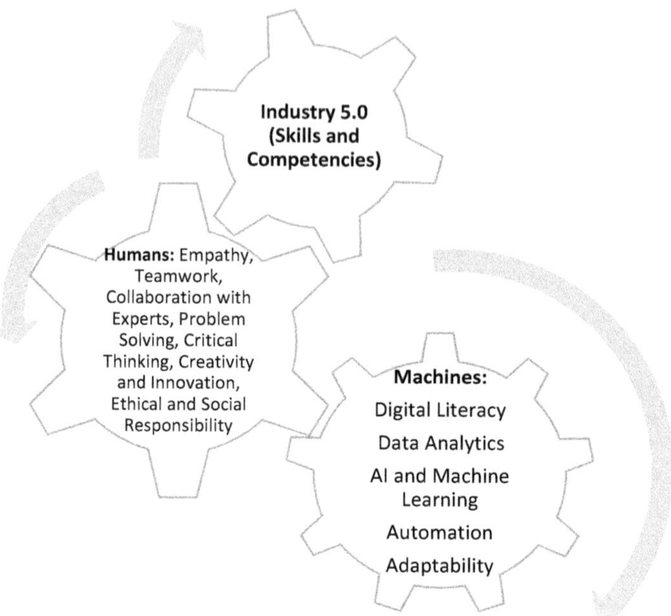

Figure 5.2 Skills and competencies demanded by Industry 5.0.

Source: Author.

5.4.2 HR's role in upskilling and reskilling the workforce

Industry 5.0 brings about a shift in the demand for skills and competencies as organizations shift to new technologies and the changing industrial landscape. This shift in the demand for skills and competencies reflects the changing demands of the workforce in Industry 5.0, where technology, data, and automation play a central role. To remain competitive, organizations and individuals need to prioritize the development of these skills which brings into picture the crucial role of HR in upskilling and reskilling the workforce. HR departments play a central role in facilitating the development and enhancement of employees' skills to meet the demands of rapidly evolving industries.

HR can adopt strategies like skill gap analysis to identify existing skill gaps in the light of current skill set of the workforce and compare it with future skills required for Industry 5.0 (Gupta, M., & Bamel, N., 2021). Anticipating future skills requirements with the emerging trends and technologies is also crucial for HR professionals. This proactive approach allows HR to plan for skills that will be essential in Industry 5.0 (World Economic

Forum, 2018. "The Future of Jobs Report") to effectively support upskilling and reskilling in Industry 5.0. HR should conduct a thorough analysis to identify existing skills gaps within the organization. This involves assessing the current skill set of the workforce and comparing it to the skills required for Industry 5.0.

One of the key skills and competencies that is increasingly in demand in Industry 5.0 is the ability to work with data and understanding of data analytics. Employees should be competent with data-driven decision-making and have proficiency in data analysis tools (Brynjolfsson & McAfee, 2014).

5.5 PERFORMANCE MANAGEMENT AND EMPLOYEE ENGAGEMENT: LEVERAGING TECHNOLOGY FOR PERFORMANCE MANAGEMENT, REAL-TIME FEEDBACK MECHANISMS IN INDUSTRY 5.0

Evolution from Industry 4.0 to Industry 5.0 emphasized the serious role of human–machine collaboration where technology has enabled better performance management and more workforce engagement. Technological advancements have changed the way organizations manage the performance of their employees and engage them. Technology enabled organizations to collect a huge amount of data and further analyze them and make data-driven decisions. The role of AI and data analytics holds importance in collecting and analyzing the data.

Real-time personalized performance metrics provides the facility for employees to monitor their progress and adapt strategies to improve their performance. Gamification knocks employees' intrinsic motivation by implementing game-like elements, such as points, badges, and leaderboards, which makes the work more engaging and enjoyable. Gamification facilitates employees to monitor their progress, set goals, and sync their efforts with the organization's objectives. This increases motivation and employee engagement as they understand how their contributions are relevant to the organizations. Employees seem to be more engaged when they are given clear targets which make them more self-responsible (Deci, L. Edward and Richard M, Ryan, 2013).

5.5.1 Fostering employee engagement in tech-driven environments

In today's tech-driven workplaces, employee engagement has become a challenging factor as the integration of technology has reshaped the way employees work. Today's workplace demands new strategies to engage employees. Fostering employee engagement in tech-driven environments includes strategic leadership and leveraging technology to create an inclusive workplace (Einsenberger et al, 1986) Effective and transparent communication enhances trust and engagement. Rise of technology has enabled adoption of remote work and flexible schedules. Providing flexible arrangements through collaborating technology and project management software facilitates communication for remote and dispersed teams. Real-time communication, document sharing, etc. has revolutionized the way teams interact. Introduction of automation and AI in routine tasks enabled employees to focus on meaningful tasks leading to higher satisfaction and engagement (Brynjolfsson & McAfee, 2014).

5.6 ETHICAL AND SOCIAL CONSIDERATIONS: ENSURING DATA PRIVACY AND SECURITY IN INDUSTRY 5.0

Data privacy is a crucial necessity for Industry 5.0 applications because the entire ecosystem uses pricey intellectual property, expensive manufacturing materials, and subscription management (Mason, Richard O,2017). To connect machines and people, designers and other collaborators, and to transmit monitoring and control information, data is exchanged over the Internet in Industry 5.0. To maintain the confidence of the cloud manufacturing ecosystem, such data must not be accessible to hostile Internet users.

Blockchain has a lot to offer Industry 5.0 in terms of security and privacy concerns. Because the blockchain uses resources, it can get slower as the number of Industry 5.0 applications using it rises. By moving the infrequently used data to the side chain, a lightweight block-chain framework can be implemented to prevent this. Additionally, without any downtime, quantum computing can be utilized to secure the cyber physical system (CPS) (Tosh et al. 2020) or cyber physical social system (CPPS) (Floridi, 2006).

5.6.1 Mitigating bias and ensuring fairness in AI applications

All models are created by people and so contain biases. The biases of organizational teams, the designers on those teams, the data scientists who use the models, and the data engineers who collect the data can all be seen in AI systems (Ng, Andrew,2017). Naturally, they also exhibit the bias that the data itself naturally possesses. To prevent and reduce unintended AI bias, teams must continue to be diverse in terms of both demographics and skill sets. To ensure AI is used fairly and without bias in HR, the following measures can be taken.

> **Diverse and Representative Training Data:** To reduce biases and prevent replicating historical disparities, AI models should be trained on extensive, varied, and representative data.
> **Regular Algorithmic Audits:** Any biased results or discriminatory trends can be found and corrected with the use of routine audits of AI

algorithms. This includes reviewing the effectiveness and effects of AI systems on various demographic groups, keeping a lookout for potential biases, and taking necessary corrective action.

Transparency: Transparency can be improved by using interpretable AI models that can explain their choices, enabling HR professionals to recognize and correct any biases that may exist. This promotes accountability and helps establish trust in the HR AI systems.

Data Privacy and Security: AI in HR is reliant on enormous amounts of employee data, which requires extreme caution. To protect employee data from unauthorized access or breaches, organizations must set up strong data governance frameworks, implement encryption and access restrictions, and adhere to pertinent data protection standards.

Compliance with Legal and Ethical Standards: AI in HR must abide by the law and ethical principles. Organizations must make sure AI systems abide by all applicable legal obligations, including non-discrimination rules and employment legislation. Additionally, AI algorithms and decision-making procedures should incorporate ethical factors like responsibility, transparency, and justice.

5.6.2 Navigating the human–machine interface

A user interface or dashboard that connects a person to a machine, system, or device is known as a human–machine interface (HMI). It can enable the automation of basic HR processes like screening resumes, responding to frequently asked queries, and scheduling interviews with previously unheard-of efficiency and precision. This automation improves the fairness and correctness of these processes by reducing the chance of human error while also saving HR personnel important time. Additionally, it is crucial in the decision-making process. It can produce actionable insights to enable better-informed HR decisions by examining patterns and forecasting trends Paul, Richard, and Linda Elder,2006). The HR experience can be customized via HMI. Understanding each person's preferences and needs enables it to customize communications and recommendations, providing a more engaging experience for both candidates and employees.

5.7 HR TECHNOLOGY IN INDUSTRY 5.0: UTILIZING HR INFORMATION SYSTEMS AND CLOUD SOLUTIONS

Organizations can use social, mobile, analytical, and cloud technologies to optimize their HR more effectively and consistently because of the shift to the digital HR paradigm. Analytics, virtual technology, and modernized learning platforms are among the HR technology priorities. To meet this need, organizations are implementing cloud-based HR systems and human resource management systems (HRMS). Along with evaluating and boosting employee productivity, teamwork, and many other areas, they enable managing a remote workforce.

HR can access real-time employee data and make informed decisions by using cloud-based HR technologies. An HRMS or HRIS (Human Resource Information System), usually referred to as a cloud-based HR system, is a web-based software solution that helps businesses manage and automate numerous HR operations. Cloud-based HRMS, in contrast to conventional on-premises HR systems, are hosted on remote servers and accessed via the Internet. This indicates that users can use a web browser or specific mobile applications to access the system at any time and from any location.

Employee-related data is processed and stored by cloud-based HR systems in a safe, centralized database. As given below, HR makes use of cloud computing and HR information systems:

Data Storage and Security: Personal information about employees, such as employment history, performance reviews, and training records, are safely recorded in the cloud. Data integrity and protection are ensured by sophisticated security methods, such as encryption and routine backups.

i.   **Employee Self-Service:** Employee self-service portals are frequently included in cloud-based HR systems, giving staff members access to their pay stubs, personal information, and business policies. Employee empowerment, decreased administrative costs, and enhanced internal communication are all benefits of this self-service functionality.

ii. **Recruitment and Applicant Tracking:** By automating processes like job posting, candidate screening, interview scheduling, and application tracking, cloud-based HR tools streamline the hiring process. Recruiters can rapidly and precisely extract pertinent data from resumes and applications using these systems' frequently available resume parsing features.

iii. **Performance Management:** With cloud-based HRMS, performance management and evaluation are made simpler. Using the system, managers may create objectives, offer ongoing feedback, and carry out performance evaluations. Managers may make informed judgments about promotions, awards, and training opportunities by tracking performance data over time.

iv. **Time and Attendance Management:** Time and attendance management tools are accessible in cloud-based HR systems, enabling workers to record their working hours, monitor overtime, and request time off. These systems frequently include mobile applications or biometric device integration for convenient time tracking.

v.   **Analytics and Reporting:** HR managers and professionals can create real-time reports on numerous HR key performance indicators

(KPIs) using HRMS's robust reporting and analytical features. These insights support businesses in trend identification, data-driven decision-making, and HR strategy optimization.

5.7.1 AI integration in HR processes

George et al (2019) with the help of AI, which has gradually saturated many HR activities, HR has been undergoing a huge disruption. Vardalier et al (2020) with the assistance of a virtual assistant, many formerly fully human-performed tasks are reconfigured. Headhunting, appraisal, training, etc. are some of the HR processes that have changed widely due to use of AI.

5.7.2 Talent acquisition

Talent acquisition is one of the monotonous jobs of HR. This process has various steps like screening resumes, evaluation based on skill set, doing preliminary interviews, tracking, following up with candidates, arranging final rounds, selection, and onboarding. With the advent of AI and websites like LinkedIn, Glassdoor, Naukri, and Monster, which are using algorithms to provide better candidates based on the candidate's profile data against the required skill set for a job. These are using keywords, search history, and connections also for data collection of a candidate. The traditional process of recruitment was posting job ads on various platforms, searching for candidates which was a time-consuming process, screening and selecting candidates' resumes was also a very tedious job, especially selecting a candidate from various job applicants. Now with digital technology, software companies can take audio or video interviews using AI, making the process quick and easy.

5.7.3 Learning and development

Learning and development process has also changed a lot due to use of AI. Learning and development program is a continuous process required at each level and frequently to increase productivity, as well as for the upgrading of the skill set of both old and new workers. With the use of AI in learning and development process, HR is able to focus on high-quality programs based on new trends and requirements, like developing a new skill that fits a vast audience. It can be made available quickly and easily to everyone using AI using different platforms.

5.7.4 Employee engagement

AI has changed the whole scenario of employee engagement, with real-time data availability like health of an employee, providing assistance virtually for any requirement, providing training of corporate policies, etc. which is again personalized. Also, AI delivers continuous feedback seamlessly and in real time, which can help in defining the KRA of each employee individually or personalized. These defined KRA and feedback data flow can give better results. This data flow from a wide audience can be managed easily using AI and can be used for making new trends and policies.

5.8 CHALLENGES AND BENEFITS

5.8.1 Challenges

   i. **Data Privacy and Security Concerns** – AI in HR raises severe concerns about data security and privacy. A breach in data security can have disastrous repercussions, including identity theft, financial losses, and reputational harm.
  ii. **Bias in AI Algorithms** – Bias in AI algorithms can be a significant problem for HR departments that use AI to make decisions. AI algorithms can be biased, just like people, leading to unjust and discriminating outcomes.
 iii. **Resistance to Change and Lack of Understanding** – AI in HR is reshaping the workplace and pushing the limits of what is achievable. However, there is always significant resistance to change.
  iv. **High Costs and Technical Expertise Required** – The cost of implementing and maintaining AI in HR can be high, and it requires technical expertise. For smaller organizations or those with tighter budgets, this can be a huge difficulty.

5.8.2 Benefits

   a. **Increased Efficiency and Productivity** – AI is being utilized in HR to automate many manual activities, allowing HR practitioners to concentrate on more strategic, high-level work while also enhancing productivity and efficiency.
   b. **Improved Candidate Experience** – The main goal of employing AI in HR is to enhance the application experience. AI significantly contributes to sustaining the time-consuming but crucial HR chores while freeing up time to address other pressing challenges.
       i. **Personalized communication:** By personalizing contact with candidates, HR professionals may improve the candidate's experience.
      ii. **Increased fairness and objectivity:** AI-powered HR technologies can reduce hiring bias, giving all candidates a more unbiased and equitable experience.

iii. **Improved scheduling:** Automation of the scheduling process with AI-powered solutions can save candidates time and effort by helping them discover a good time for an interview.

iv. **Automated feedback:** AI-enabled platforms can give candidates automated feedback on the status of their applications, cutting down on the amount of time they have to wait for a response from HR

v. **VR experiences:** To help candidates better understand the firm and enhance the candidate experience, organizations use AI-powered VR experiences to take them on an investigation of the workplace environment.

vi. **Chatbots:** Chatbots powered by AI may respond to frequent candidate inquiries around the clock, making the hiring process more streamlined and approachable.

c. **Enhanced Data Analytics and Insights** – Move over, spreadsheets – AI is advancing on the HR stronghold with an abundance of improved data analyses and insights.

d. **Better Decision-Making** – AI's powerful algorithms and machine learning prowess are opening new avenues for HR data and insights, offering decision-makers access to an immense amount of data.

e. **Increased Fairness and Objectivity** – AI has the potential to eliminate human biases and preconceptions that might infiltrate HR decision-making because of its capacity to evaluate massive volumes of data and generate objective suggestions.

f. **Better Work Culture** – With AI, HR can create an environment at work where the efficiency and accuracy of technology meet the humanity and empathy of interpersonal contact.

5.9 FUTURE PERSPECTIVES: HR IN AN EVER-EVOLVING INDUSTRY 5.0

5.9.1 Embrace Industry 5.0 technologies

Industry 5.0 likely involves even greater integration of technologies, such as the IoT, AR, and advanced robotics. HR needs to understand these technologies and their implications for the workforce. This might involve recruiting individuals with expertise in these areas or upskilling existing employees. Also, traditional workforce planning may become obsolete in the face of rapid technological changes. HR should adopt agile workforce planning methodologies, allowing for quick adjustments in response to technological advancements. This involves identifying future skills, planning for reskilling, and upskilling, and creating a flexible workforce. Industry 5.0 may also involve global collaboration and diverse teams. HR should foster a culture of adaptability and inclusivity. This includes promoting cultural intelligence, understanding the nuances of working with international teams, and creating policies that respect diverse cultural norms. In addition to this, the upcoming HR fraternity may witness increased mobility of talent across functions and geographical locations. Therefore, HR needs to develop strategies for workforce mobility, ensuring that employees have opportunities to develop diverse skill sets and contribute to various parts of the organization.

5.9.2 Human-centric technology implementation

With the rise of automation and AI, HR will need to focus on creating a harmonious collaboration between humans and machines. This includes designing roles that complement automation, providing training for employees to work alongside AI, and ensuring that technology enhances rather than replaces human capabilities. Along with this, HR professionals must ensure that technology implementation remains human-centric. While automation and AI are powerful tools, they should align with human values and ethical considerations. HR should be involved in the ethical use of technology, promoting fairness, transparency, and accountability.

5.9.3 Advanced learning and development programs

HR professionals must attempt to develop advanced learning and development programs that are responsive to Industry 5.0 requirements. They must use technologies such as VR and AR to provide immersive training experiences.

5.9.4 Digital leadership development

The HR of the organization should make an attempt to cultivate digital leadership skills within the organization. Also, leaders will need to understand and harness the potential of emerging technologies, lead digital transformation initiatives, and navigate the complexities of a technologically advanced environment. HR can play a key role in identifying and developing digital leadership capabilities.

5.9.5 Data-driven HR decision-making and cybersecurity issues

In Industry 5.0, data will be abundant. HR professionals should be adept at using data analytics for decision-making. This includes predictive analytics for talent acquisition, performance management, and identifying areas for workforce improvement. Data-driven insights will be crucial for strategic HR planning. As technology becomes more integrated into HR processes, there is an increased need for robust cybersecurity measures. HR should collaborate with IT to ensure the security of employee data, especially as more HR functions move to cloud-based platforms and AI applications.

5.9.6 Flexible work arrangements and remote collaboration

Given the advancements in connectivity and communication technologies, HR needs to refine policies and practices related to flexible work arrangements and remote collaboration. This includes creating a seamless virtual work environment, managing distributed teams effectively, and ensuring the well-being of remote workers. In Industry 5.0, the employee experience extends beyond the physical workplace. HR should focus on creating a holistic employee experience that encompasses both physical and virtual dimensions. This involves considering the impact of technology on employee engagement, satisfaction, and well-being.

5.10 FINDINGS

The growing integration of robots into enterprises will provide new challenges for HR departments. Along with their present responsibilities, they will also need to decide which tasks robots will replace. It basically means that they will decide which tasks to delegate to robots.

Applying digital HRM techniques has made it easier to choose which candidates to recruit, analyze the skill requirements needed to fill open positions, and choose which prospects to select. The development of modern digital technology has made integration and flexibility possible. Training and development services are undergoing a profound transformation thanks to digital technologies. With employee behavior being more and more logged on data platforms, these assessments are growing in precision nowadays. Corporations may develop a database with details regarding employee performance critiques, tardiness, staff turnover, financial potential for the company, and other subjects. Transitioning to remote work is one of the biggest changes for HR. Managing and engaging employees remotely is the responsibility of HR specialists. The way organizations interact and oversee employee performance has evolved as a consequence of technological advancements. Gamification is an approach that promotes employee engagement and enjoyment by introducing elements of games, such as leaderboards, badges, and points, to undermine employees' intrinsic motivation. Industry 5.0 applications require data privacy because the whole ecosystem relies on expensive manufacturing materials, intellectual property, and subscription management.

HR can use cloud-based HR technologies to access real-time employee data and make informed decisions. Often called a cloud-based HR system, an HRMS or HRIS is a web-based software program that assists companies in managing and automating various HR procedures.

5.11 IMPLICATIONS AND RECOMMENDATIONS

Industry 5.0 is characterized by swift technological advancements, requiring HR professionals to stay abreast of emerging technologies and understand their impact on the workforce. Also, the skills required in Industry 5.0 may change rapidly; therefore, HR needs to identify the evolving skill sets and work on strategies for upskilling and reskilling the workforce accordingly. The time to come promotes an integration of AI and robotics which implies a need for HR to manage the collaboration between humans and machines, ensuring a harmonious and productive working relationship. Also, it is expected that traditional workforce planning may not suffice in the face of rapid changes; therefore, HR needs to adopt agile methodologies for workforce planning to respond quickly to evolving industry needs. This includes creating cross-functional teams, fostering a culture of learning and experimentation, and facilitating easy transitions between roles. Offering cross-cultural training programs will help employees and HR professionals in navigating the challenges of working in a globalized Industry 5.0. This includes understanding cultural nuances, effective communication across cultures, and promoting an inclusive workplace culture.

In addition to this, an abundance of data in Industry 5.0 necessitates a shift towards data-driven decision-making in HR processes, from talent acquisition to performance management. With the possibility of widespread remote work, HR must address challenges related to virtual team management, employee engagement in a virtual environment, and well-being support for remote workers. Industry 5.0 may involve a more globally distributed workforce. HR needs to manage the complexities of global talent management, including cultural differences, legal considerations, and collaboration across time zones. HR must focus on developing leaders who can navigate the digital landscape, lead digital transformation initiatives, and effectively manage teams in a technologically advanced environment.

HR professionals are required to establish robust continuous learning programs to ensure that their employees stay updated on evolving technologies. For this, they can make use of online courses, workshops, and partnerships with educational institutions for ongoing skill development. This will empower HR teams to leverage data for strategic decision-making and enable employees to understand and use data in their roles. There is a need to make a revision in the recruitment strategies to identify candidates with not only the required technical skills but also the ability to adapt to rapid technological changes. They can also use AI in recruitment processes to enhance efficiency. Apart from this, investment in advanced virtual collaboration tools is crucial. This primarily includes video-conferencing platforms, project management tools, and VR applications for immersive collaboration.

Reviewing and adapting HR policies to accommodate the flexibility needed in Industry 5.0 is the need of the hour. This includes policies related to remote work, flexible hours, and benefits that support the diverse needs of the workforce. Prioritizing employee well-being initiatives, especially in the context of remote work, is equally important. Providing resources for mental health support, stress management, and encouraging a healthy work–life balance is very basic in the Industry 5.0 era. Another major area of concern is the fostering of a strong collaboration between HR and IT teams to ensure the security of employee data in an increasingly digital HR environment. Implementing cybersecurity measures and best practices to protect sensitive information is needed at every step. Establishing clear ethical guidelines for the use of technology in HR processes is crucial. This will include AI in recruitment, employee monitoring technologies, and other applications. Communication of these guidelines transparently is another crucial factor in building trust among employees.

REFERENCES

Amabile, Teresa M. "The case for the social psychology for creativity" in *Creativity in context: Update to the social psychology of creativity*, edited by Teresa M. Amabile, 15–21 Routledge, 2019.

Anderson, Craig A., and Karen E. Dill. "Video games and aggressive thoughts, feelings, and behavior in the laboratory and in life." *Journal of Personality and Social Psychology* 78, no. 4 (2000): 772.

Betchoo, Nirmal Kumar. "Digital transformation and its impact on human resource management: A case analysis of two unrelated businesses in the Mauritian public service." in *2016 IEEE international conference on emerging technologies and innovative business practices for the transformation of societies (EmergiTech)*, pp. 147–152. IEEE, 2016. https://doi.org/10.1109/EmergiTech.2016.7737 328, https://ieeexplore.ieee.org/abstract/document/7737328

Brynjolfsson, Erik, and Andrew McAfee. "The skills of the new machines: Technology races ahead" in *The second machine age: Work, progress, and prosperity in a time of brilliant technologies*, edited by Brynjolfsson Erik, McAfee Andrew, 13–38 WW Norton & Company, 2014.

Deci, Edward L., and Richard M. Ryan. *Intrinsic motivation and self-determination in human behavior*. Springer Science & Business Media, 2013.

DiRomualdo, Anthony, Dorothée El-Khoury, and Franco Girimonte. "HR in the digital age: How digital technology will change HR's organization structure, processes and roles." *Strategic HR Review* 17, no. 5 (2018): 234–242.

Eisenberger, Robert, Robin Huntington, Steven Hutchison, and Debora Sowa. "Perceived organizational support." *Journal of Applied Psychology* 71, no. 3 (1986): 500.

Fenech, Roberta. "Human resource management in a digital era through the lens of next generation human resource managers." *Journal of Management Information & Decision Sciences* 25 (2022).

Friedman, Batya, and David G. Hendry. "Introduction" in *Value sensitive design: Shaping technology with moral imagination* by Friedman Batya and David G Henry 1-19, MIT Press, 2019.

Floridi, Luciano. "Four challenges for a theory of informational privacy." *Ethics and Information Technology* 8 (2006): 109–119.

Grant, A. M., Fried, Y., & Juillerat, T. (2011). Work matters: Job design in classic and contemporary perspectives. In *APA handbook of industrial and organizational psychology, Vol. 1. Building and developing the organization* edited by Sheldon Zedeck, 417–453). Washington, DC: American Psychological Association, 2011. https://doi.org/10.1037/12169-013

Gupta, M., & N. Bamel. "A systematic review on skill gap analysis," *Journal of Critical Reviews* 7 (2021).

Maddikunta, Praveen Kumar Reddy, Quoc-Viet Pham, B. Prabadevi, Natarajan Deepa, Kapal Dev, Thippa Reddy Gadekallu, Rukhsana Ruby, and Madhusanka Liyanage. "Industry 5.0: A survey on enabling technologies and potential applications." *Journal of Industrial Information Integration* 26 (2022): 100257.

Mason, Richard O. "Four ethical issues of the information age." In *Computer ethics*, pp. 41–48. Routledge, 2017.

Nahavandi, Saeid. "Industry 5.0 – a human-centric solution." *Sustainability* 11, no. 16 (2019): 4371. https://doi.org/10.3390/su11164371

Ng, Andrew. "Artificial intelligence is the new electricity." In *Presentation at the Stanford MSx future forum*. 2017.

Paul, Richard, and Linda Elder. "Critical thinking: The nature of critical and creative thought." *Journal of Developmental Education* 30, no. 2 (2006): 34.

Sindhwani, Rahul, Shayan Afridi, Anil Kumar, Audrius Banaitis, Sunil Luthra, and Punj Lata Singh. "Can Industry 5.0 revolutionize the wave of resilience and social value creation? A multi-criteria framework to analyze enablers." *Technology in Society* 68 (2022): 101887.

Schwab, Klaus. *The fourth industrial revolution*. Crown Currency, 2017.

Tosh, Deepak, Oscar Galindo, Vladik Kreinovich, and Olga Kosheleva. "Towards security of cyber-physical systems using quantum computing algorithms." In *2020 IEEE 15th international conference of system of systems engineering (SoSE)*, pp. 313–320. IEEE, 2020.

Vardarlier, Pelin, and Cem Zafer. "Use of artificial intelligence as business strategy in recruitment process and social perspective." In *Digital business strategies in blockchain ecosystems: Transformational design and future of global business*, pp. 355–373. 2020.

Vladyslav, Y., and G. Dmytro. "Comparative Analysis of the Basic Methods Used in Industry 4.0 and Industry 5.0." PhD diss., 2023.

Additional Reading

Aslam, Farhan, Wang Aimin, Mingze Li, and Khaliq Ur Rehman. "Innovation in the era of IoT and Industry 5.0: Absolute innovation management (AIM) framework." *Information* 11, no. 2 (2020): 124.

Demir, Kadir Alpaslan, Gözde Döven, and Bülent Sezen. "Industry 5.0 and human—robot co-working." *Procedia Computer Science* 158 (2019): 688–695.

George, Ginu, and Mary Rani Thomas. "Integration of artificial intelligence in human resource." *International Journal of Innovative Technology and Exploring Engineering* 9, no. 2 (2019): 5069–5073.

Chapter 6

E-customer safety in digital environment from the seller's and the buyer's perspectives

*Shashi Kant Gupta, Joanna Rosak-Szyrocka,
V. Suresh Kumar, and Gilbert C. Magulod Jr.*

6.1 INTRODUCTION

In the digital age, the idea of e-customer safety has grown in importance as a value determinant, influencing the nature of correspondence and payments. As the internet continues to permeate every part of contemporary life, companies, consumers, and governments alike are becoming concerned about maintaining the privacy and security of online transactions and communication (Alazzam, Shakhatreh, Gharaibeh, Didiuk, and Sylkin 2023).

E-customer safety is not a basic requirement in today's digital environment, but it also plays a significant role in fostering customer confidence and loyalty. Businesses' reputation as a whole and level of competition are directly impacted by their capacity to ensure the security of their customers' confidential data. As a growing number of high-profile hacking attempts and data breaches make news, consumers are providing personal internet with more caution and discernment (Lova and Budaya 2023).

The concept of e-customer safety has become extremely important in the changing world of digital business, especially when it comes to cybersecurity. As the intensity and sophistication of cyber-attacks continue to rise, protecting online payments and customer data has become a critical factor in determining the value of enterprises engaged with online commerce. The complexity of safety has become essential to the authenticity of electronic transactions, requiring a thorough comprehension of the precautions needed to protect sensitive information from hostile activities like hacking, stealing of identities, and data breakdowns (Kumar, Singh, Mohanty, Goel, Gupta, Alharbi, and Khanna 2023).

The increasing frequency of crime has highlighted the necessity for companies to give priority to robust safety policies, not only as a response to possible attacks but also as an essential part of their operational plan. The implications of hacking into computers are extensive and extend much further than financial losses; it frequently destroys the brand of an organization and undermines the confidence of its clientele. As a result, companies are forced to spend a lot of money on cutting-edge cybersecurity knowledge

　　　　　　　　　DOI: 10.1201/9781032677040-6

and tools to proactively identify, stop, and lessen the risks associated with bad actors trying to take advantage of holes in electronic systems (Nalluri and Rao 2023).

In the context of e-commerce, e-customer safety is the critical nexus of electronic transactions and the field of cybersecurity, emphasizing the necessity of protecting online interactions with customers and information privacy. With the ever-expanding realm of online purchasing, it is imperative that technological infrastructure be strengthened against ever-evolving security risks. The key to creating secure e-commerce surroundings that inspire confidence in customers and help businesses compete in the online marketplace is to ensure that strong cybersecurity measures, secret payment processors, and strict safeguarding of information procedures are in place (Khan, Ansari, Ansari, and Amir 2023). Figure 6.1 represents the different types of threads pertaining to customer security.

This study aims to emphasize the necessity of long-term fixes to improve customer protection in the e-commerce industry by identifying risks and presenting feasible alternatives to support online safety and advance a secure digital marketplace.

The remaining portion of this article is as follows: In Part 6.2 of this article, we provide the related work. Part 6.3 discusses the methodology. In Part 6.4 of this article, we provide the result and discussion. The concluding segment

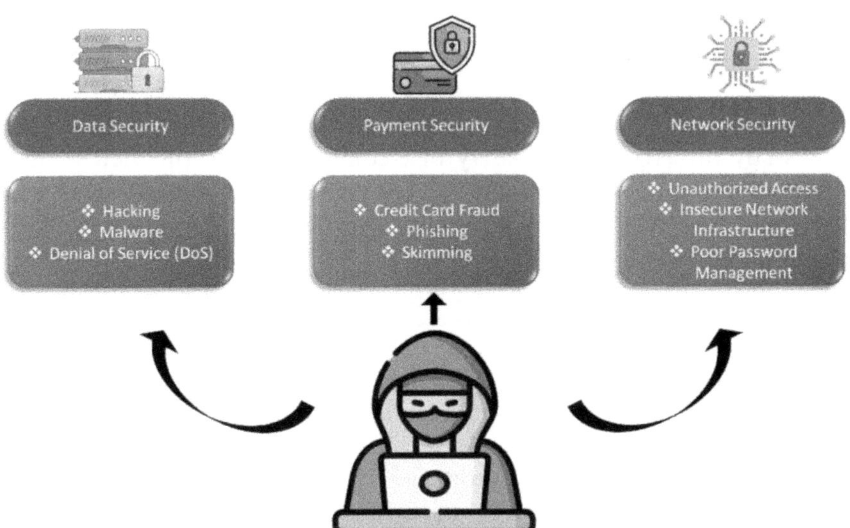

Figure 6.1 Different types of threads pertaining to customer security.

Source: Author.

of this article, Part 6.5, summarizes the key findings and contributions of our research.

6.2 RELATED WORK

A study (Yoo, Lee, and Atamja 2023) examined the relationship between consumer fulfillment and e-commerce data quality and design of websites and how user purchasing commitment in South Korea was influenced by consumer happiness. Knowledge of the antecedent elements that cause Indian customers to have concerns when utilizing e-commerce websites was presented in the study by Pramanik and Prabhu (2022). The study by Sopyan, Febrian, and Sani (2023) evaluated Indonesian consumers' levels of electronic happiness and commitment with regard to online purchasing and sales platforms. The study by Huy and Phuc (2023) produced a comprehensive and accurate examination of the ways in which intelligence services and large amounts of data as a service offer small and medium-sized businesses a number of avenues for success in e-commerce. A study by Prabhavathy, Senthilkumar, Subathra, Jinu, and Kumar (2023) determined the way value for customers, e-satisfaction, e-trust, and quality of e-services impact Indonesian online shopping users' e-loyalty. The study by Apriansyah and Yacob (2023) examined the connections between e-customer interaction, attitudes regarding online purchasing, and the experience of purchasing online. The impact of service quality and service loyalty on e-commerce companies was conducted in the study by Al-Khateeb, Jaoua, and Mohamed (2023).

6.3 METHODOLOGY

This study investigated the safety of e-commerce from the viewpoints of online shoppers and online retailers. The more the dangers from the seller and the customer, the less adequate security. The following were the research's primary objectives:

- Determining which hazards e-customers are not aware of and which they view as real.
- Analyzing the fear of customers shopping online.
- Identifying risks to the operation of the online store that affects client security.
- Presenting mitigation strategies that should promote sustainable development.

Two systems were used as the foundation for the organized and reasonably thorough threat identification process. Regarding online shoppers, the dangers found in the various stages of the buying process were organized.

Table 6.1 Threats that clients face during different stages of an e-commerce

Steps associated with purchasing things online	Risks associated with specific online shopping stages
Product Selection	• Phishing Website • False Product Information • Manipulate Pricing Strategies
Adding to Cart and Checkout Process	• Unauthorized Access to Shopping Cart • Data Interception
Payment Authorization	• Payment Card Fraud • Man in the Middle Attack • False Translation Confirmation
Order Processing and Fulfillment	• Delivery Delays and Non-Delivery • Product Quality Control

Source: Author.

Every stage embodies the distinct risks that consumers face when they shop online, emphasizing the obstacles including identity theft, unsafe payment options, delivery issues, and fraudulent schemes. Table 6.1 represents the threats that clients face during different stages of e-commerce.

Several steps are involved in the online purchasing process, starting with product selection, where risks such as encountering phishing websites, misleading product information, and manipulated pricing methods can pose threats to consumers. Moving to the purchase procedure, potential risks include unauthorized access to the shopping cart, leading to unauthorized changes or disclosure of personal information, as well as the interception of sensitive data during the checkout process. In the payment authorization step, risks include payment card fraud, man-in-the-middle attacks, and false translation confirmation, all posing threats to the security of financial transactions. In the final phase of order processing and fulfillment, potential risks involve delivery delays, which can lead to customer dissatisfaction, and the importance of maintaining product quality control to ensure customer happiness and prevent post-sale issues.

6.4 RESULT AND DISCUSSION

We address the safety of e-commerce consumers in the first section of the empirical results and the safety of e-stores in the second. Table 6.2 and Figure 6.2 present data regarding the relationship between the number of threads and the percentage of threads that have an impact on people.

The product selection stage reflects an 83% impact on individuals, emphasizing the challenges associated with choosing items online. Moving to the cart and checkout process stage, 85% of the difficulties pertain to people, highlighting potential hurdles during the checkout process. Payment

Table 6.2 Proportion of threads that impact people

Threads	Percentage of threads to affect people
Product Selection	83
Adding to Cart and Checkout Process	85
Payment Authorization	95
Order Processing and Fulfillment	90

Source: Author.

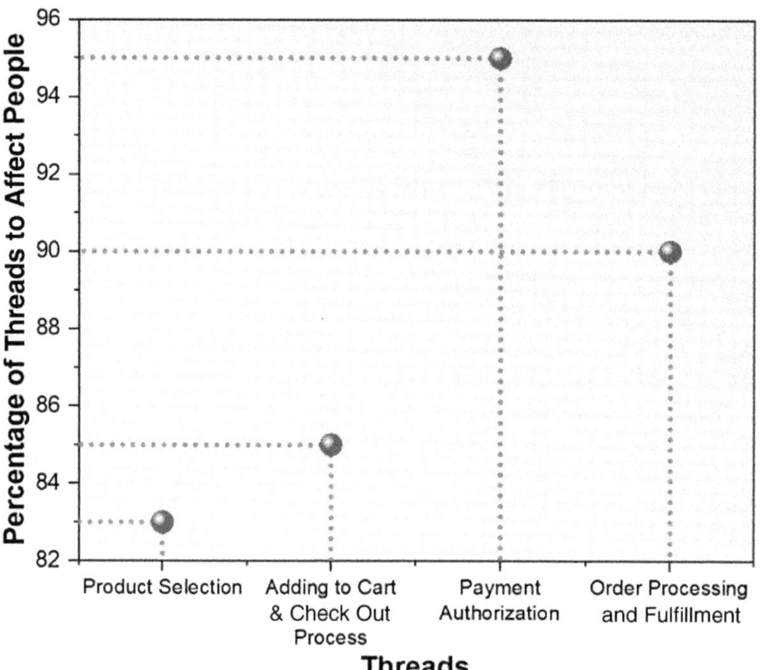

Figure 6.2 Threads that impact people.

Source: Author.

authorization exhibits a significant impact of 95%, suggesting concerns regarding the safety of payment information. Finally, the order processing and fulfillment stage demonstrates a 90% impact, underscoring challenges related to order completion and timely delivery. Table 6.3 and Figure 6.3 represent the possibilities of eliminating ways to improve online security.

Table 6.3 Possibilities of eliminating ways

Eliminating ways	Possibility
Implementing Multi-factor Authentication	80
Regular Security Audits and Updates	75
Enhancing Customer Awareness	62
Ensuring Transparency in Product Listings	65
Investing in Prevent System	50

Source: Author.

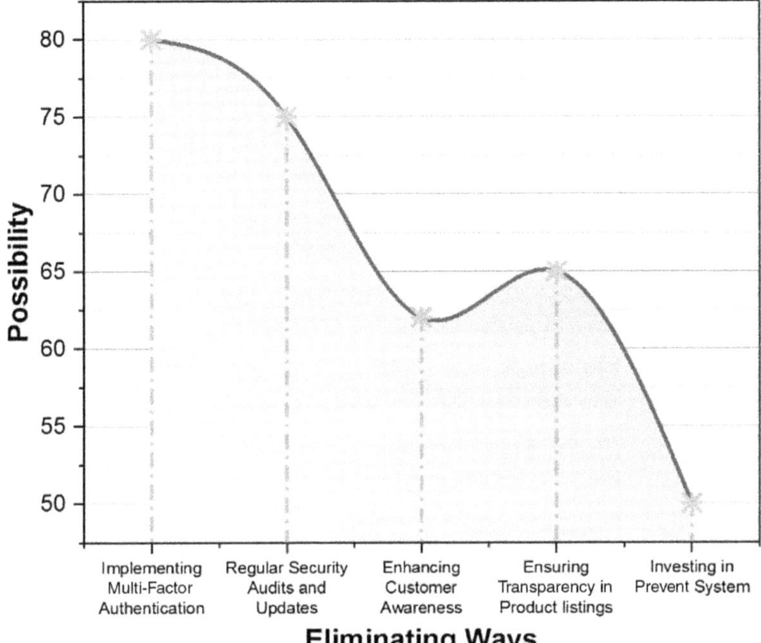

Figure 6.3 Opportunities to get rid of techniques.

Source: Author.

The potential effectiveness of approaches for enhancing online security is evident, with multi-factor authentication ranking highest at 80%, followed by regular security audits and updates at 75%. Enhancing customer awareness and ensuring transparency in product listings hold possibility ratings of 62% and 65%, respectively, while investing in preventive systems

Table 6.4 Customer concerns

Fear of customer	Percentage of effects
Delivery and Shipping Issues	55
Data Security and Privacy	74
Reliability of Product Quality	76
Cybersecurity and Fraudulent Activities	78
Return and Refund Policies	66
Customer Support and Communication	61
Lack of Physical Examination	58

Source: Author.

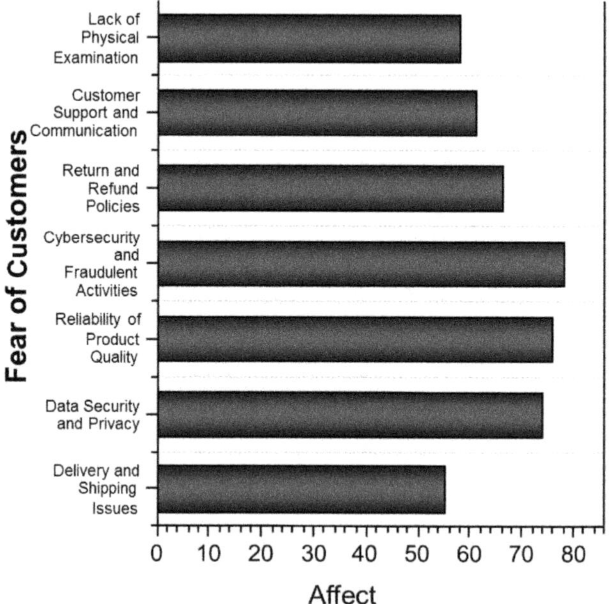

Figure 6.4 Performance of customer concerns and their impact percentages.

Source: Author.

has a moderate potential of 50%. Table 6.4 and Figure 6.4 display customer concerns and their impact percentages.

Table 6.4 highlights common consumer concerns and their impact percentages, ranging from 55% for delivery and shipping issues to 78% for cybersecurity and fraudulent activities. The data emphasizes the significance

Table 6.5 Factors that lead to the development of risks to online retailers' ability to operate

E-store elements	Reasons for the emergence of threats
Staff	– Absence of cybersecurity experts on staff – Insufficient cybersecurity training and self-education – Employees having unrestricted or inadequate access to private information – Not adhering to website access along with app protocols
IT equipment	– Older components for computers – Insufficient security measures for hardware used for computer access
Software	– Absence of cyberattack software for detection – Absence of advanced antivirus software – Absence of software or neglecting to do the testing to find holes in the network – Insufficient or inadequate data backup security – Absence of supplementary security measures, such as two-factor authentication (2FA), for vital systems – No upgrades to software

Source: Author.

of timely delivery, data security, product reliability, online safety, return policies, customer support, and the limitations of online product examination for customers. Table 6.5 represents the reasons behind the rise of risks to online retailers' ability to operate.

The study discusses the causes of risks outlined in Table 6.5 and proposes countermeasures, emphasizing the importance of dedicated cybersecurity personnel equipped with appropriate tools. The post-breach analysis is vital for understanding the implications of cyberattacks, ensuring improved e-commerce security, and preventing future incidents.

6.5 CONCLUSION

Billions of people worldwide are impacted by the problem of customer security in the e-commerce industry, which affects a multitude of online shopping platforms. Nevertheless, the strategy used to address this problem frequently lacks sustainability, exposing gaps that fraudsters can easily exploit. This chapter highlights the crucial areas that e-customers should be concerned about, with a particular emphasis on payment card security, personal data security, payment methods security, and product security. Furthermore, the security protocols that e-commerce platforms employ are inextricably linked to the safety of their customers. The study emphasizes the way urgent need to improve cybersecurity safeguards and have an awareness of the risks. Through the prioritization of vulnerability elimination, the e-commerce industry can promote sustainable development and increase consumer confidence when making purchases online. It is limited by its focus on current e-commerce safety concerns from the consumer and seller perspectives, potentially excluding new dangers and other participants. In addition, the changing dynamics of digital security and locale-specific aspects may not be fully obtained, thus reducing the investigation's comprehensive evaluation. Future studies should concentrate on creating comprehensive and long-lasting security frameworks for the e-commerce industry, with a focus on sophisticated technologies, cooperation among regulatory agencies and all parties involved, and instruction.

REFERENCES

Al-Khateeb, Bilal Ahmad Ali, Fakher Moncef Jaoua, and Elsayed Sobhy Ahmed Mohamed. (2023). "The impact of attitude towards online shopping in strengthening the relationship between online shopping experience and e-customer engagement." *International Journal of Customer Relationship Marketing and Management (IJCRMM)* 14, no. 1: 1–25.

Alazzam, Farouq Ahmad Faleh, Hisham Jadallah Mansour Shakhatreh, Zaid Ibrahim Yousef Gharaibeh, Iryna Didiuk, and Oleksandr Sylkin. (2023). "Developing an information model for e-commerce platforms: A study on modern socio-economic systems in the context of global digitalization and legal compliance." *International Information and Engineering Technology Association* 28, no. 4: 969–974. http://iieta.org/journals/isi

Apriansyah, Roky, and Syahmardi Yacob. (2023). "Could customer value support e-commerce practices in Indonesia." *Tec Empresarial* 18, no. 1: 415–429.

Huy, Pham Quang, and Vu Kien Phuc. (2023). "Big data in relation with business intelligence capabilities and e-commerce during COVID-19 pandemic in accountant's perspective." *Future Business Journal* 9, no. 1: 40.

Khan, Amir, Saghir Ahmad Ansari, Shoaib Ansari, and Mohammad Amir. (2023). "Consumer behavior towards E-commerce during pre and post-Covid-19 in Aligarh: A comparative analysis." *Asian Journal of Economics, Finance and Management* 9: 97–103.

Kumar, Rakesh, Tilottama Singh, Sachi Nandan Mohanty, Richa Goel, Deepak Gupta, Meshal Alharbi, and Rupa Khanna. (2023). "Study on online payments and e-commerce with SOR model." *International Journal of Retail & Distribution Management* 6: 1–13.

Lova, Anggil Nopra, and Indra Budaya. (2023). "Behavioral of customer loyalty on e-commerce: The mediating effect of e-satisfaction in Tiktok shop." *Journal of Scientific Research, Education, and Technology (JSRET)* 2, no. 1: 61–73.

Nalluri, Padmaja, and R. Srinivasa Rao. (2023). "Impact of product line on shopping motivations of consumers in e-commerce – a correspondence analysis." *Global Business Review*, 09721509231170469.

Prabhavathy, R., S. Senthilkumar, K. Subathra, Rohan Thomas Jinu, and Paul Arun Kumar. (2023). "Impact of service quality on service loyalty in e-commerce firms." *Journal of Scientific Research and Technology* 1: 12–16.

Pramanik, Richa, and Sandeep Prabhu. (2022). "Analysing cyber security and data privacy models for decision making among Indian consumers in an e-commerce environment." In *2022 International Conference on Decision Aid Sciences and Applications (DASA)*, pp. 735–739. IEEE.

Sopyan, A., Febrian, W.D. and Sani, I. (2023). "Strategy to increase e-customer loyalty through e-customer satisfaction in e-commerce business in Indonesia." *Scandinavian Journal of Information Systems* 35, no.1: 38–46.

Yoo, Sungjoon, Dong-Joo Lee, and Louis Atamja. (2023). "Influence of online information quality and website design on user shopping loyalty in the context of e-commerce shopping malls in Korea." *Sustainability* 15, no. 4: 3560.

Chapter 7

How multi-agent technology ensures total quality management in organizations

Shubham Sharma and Somesh Kumar Sharma

7.1 INTRODUCTION: MULTI-AGENT TECHNOLOGY

The manufacturing industry has seen a considerable change over the past 10 years because of the advancement of numerous technologies in the field of computer engineering. The use of high-performance sensors, processing, and networking devices is responsible for this change. "Smart manufacturing" is a phrase frequently used to describe this new paradigm. The manufacturers have been helped by this paradigm to weather market volatility such as competition from developing markets and the demand for mass customization.

Agents are cognitive beings that display autonomy, responsiveness, proactivity, and social aptitude. The definition of an agent that is most frequently used is "an encapsulated computational system" that is positioned in some environment and that is able to engage in flexible, independent activities to carry out its design intentions in such setting. When dealing with complicated and significant issues like industrial systems, a single agent has its computational and knowledge constraints. One approach to solving these issues by using the aid of a Multi-Agent Technology (MAT), in which each agent utilizes its expertise to address its own unique issue while coordinating with other agents to address related issues (Pulikottil et al. 2021). Hence, the function of the agent-based system is shown in Figure 7.1.

Agents in this system cooperate by sharing data while operating autonomously. If one agent is unable to complete a job, the system continues to function properly, and another agent is automatically assigned to complete the work. As a result, MAT is a collection of unique agents that cooperate to solve issues that other methods are unable to handle (Singh and Sharma 2023). Multiple interacting intelligent agents make up the computer-based ecosystem known as MAT. It is preferable to employ MAT to resolve issues that are challenging (or impossible) for a single agent. No categorical definition of multi-agent technology exists, just like there is not one for agents, therefore let's concentrate on one that is mostly agreed upon. Also, MAT is described as a loosely linked network of things (agents) that solve problems

DOI: 10.1201/9781032677040-7

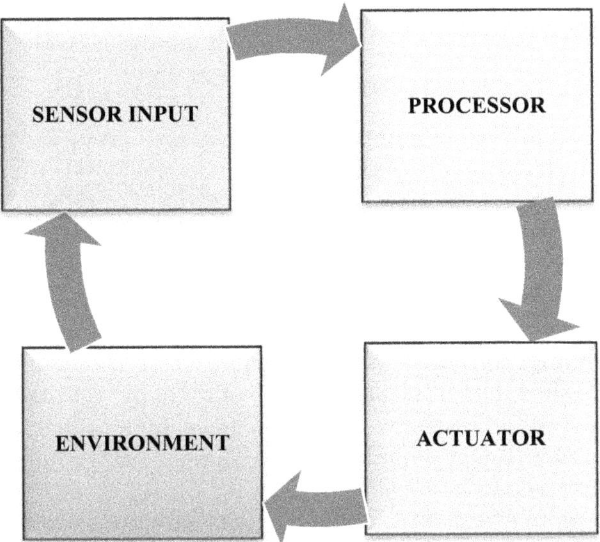

Figure 7.1 Agent-based system.

and collaborate to discover solutions to challenges that are outside the scope of each entity's particular skills or expertise. Agents in an environment interact with one another in Agent-Based Models (ABM), like MAT. Due to its many uses, which might relate to utterly distinct techniques, ABM is known by several names. It is also known as an Agent-Based System (ABS) or a Multi-Agent System (MAS) (Rocha, Boavida-Portugal, and Gomes 2017a).

7.2 WHY MULTI-AGENT TECHNOLOGY?

As the size keeps growing, modelling and calculation duties get significantly more difficult. As a result, employing centralized techniques to manage it is time-consuming and challenging. Therefore, the benefits of employing MATs are (1) individuals that take into consideration the application-specific nature and environment; (2) the modelling and investigation of local interactions between persons; and (3) the organization of modelling and computing challenges as sublayers or components. Although the motivations for using multi-agent technology for academics from diverse fields vary, the above are the main benefits of using MAT. As a result, MAT offers a useful computational paradigm for distributed control. Artificial intelligence (AI) methods can also be applied, in which a system of computers that is positioned during a situation and can act autonomously inside this setting to achieve its structural goals is referred to as an agent (Xie and Liu 2017).

7.3 MULTI-AGENT TECHNOLOGY CHARACTERISTICS

Applications may refer to various components as they are laid up in various paradigms and have varying goals (Franklin and Graesser 1997). An agent can be viewed as a component of a programme (such as a system, approach, or subsystem) or as any form of autonomous body (such as a company or a person). Each agent has a set of preprogrammed responses to other agents and its computing environment, ranging from simple decision-making in response to sophisticated AI that responds. Even so, one could assume that Wooldridge and Jennings (1995) should be supported by the majority of scholars. They established the following characteristics of an agent as a component of a computer either hardware or software as shown in Figure 7.2.

Reactivity: Refers to an agent's ability to perceive their surroundings and act swiftly in response to potential changes.

Proactivity: The ability of agents to make the first move and exhibit goal-driven behaviours rather than just reacting to the environment.

Social Abilities: By using a specific Language of Agent Communication (LAC) and making ties between their autonomy goals and the geographical

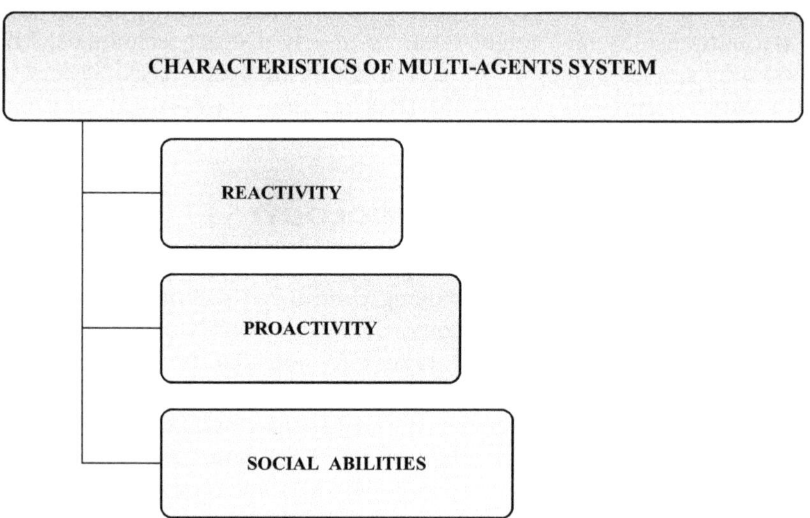

Figure 7.2 Characteristics of multi-agents.

Source: Based on Wooldridge and Jennings 1995.

surroundings. The agents may interact/communicate via other agents as well and perhaps humans (cognitive model).

These characteristics are more challenging to pinpoint than they may initially appear to be. Even though it is valued in the agent community and is a key component of Wooldridge and Jennings' definition of an agent, autonomy can never be fully attained. It is obvious that a human (or another agent) must design and implement the agent before it can begin to function. It is also not totally true to assume that the agent's activity will never come to end. Of course, the agent is going to have a finite lifespan and a last step within the auspices of present science. However, even if the agent needs autonomy – the ability to function without human intervention – interaction with humans is typically desired or even necessary. It is common practice to create autonomous agents that can also follow commands from humans. It can be easy to construct fully reactive agents, but this is not always desired. A fully reactive agent would behave in a goal-oriented manner by responding to environmental changes repeatedly without attempting to meet its medium- or long-term objectives. Agents who can balance proactive and reactive behaviour should be defined. However, striking a balance between these two behavioural styles is quite difficult (Reis, Lau, and Oliveira 2001).

In functional systems, proactiveness is straightforward to acquire. However, its straightforwardness only holds true if fixed surroundings are considered, meaning one in which nothing changes while a certain method or function is being carried out. In addition, there should be no environmental uncertainty while the agent uses that method or performs that duty these presumptions are not true in most contexts, though. Agents must be capable of responding to environmental transformations and determining whether the original goals are still met and relevant in dynamic and partially accessible surroundings while carrying out a particular operation. As a result, the agents must be reactive and capable of speedy adaptation to environmental phase changes.

An agent's social capacity is determined by its capacity to interact with other agents and/or humans in ways resembling how people interact with one another every day, exchanging high-level messages rather than just data bytes with no apparent meaning and establishing social norms. These procedures include collaboration, discussion, and coordination. It is vital to consider the goals in comparison to the other agents (if any exist in the surroundings) or at the very least have an idea that agents exist to carry out the idea.

It is also important to realize that they are independent agents with perhaps different objectives. As a result, it could be essential to bargain and work together with other actors before finally trading products or information. For instance, it can be essential to make a payment or provide a certain commodity or service to persuade an agent to collaborate. In several situations, agents are unable to participate in any cooperative process because their goals are incompatible.

It is also crucial to maintain a balance between social aptitude and proactive or reactive talents. In a cooperative work plan established by a group of agents with a shared objective, its significance is much more important. Every agent must adjust their response under these circumstances to the occurrences that take place in the modelling surroundings, using both the ability to choose and carry out tasks and their public behaviours to carry out group tasks (Reis, Lau, and Oliveira 2001).

Certain scholars have highlighted other facets of agents, such as mobility and flexibility. Agents may, in fact, have supplemental characteristics, some of which may be more important than others depending on the situation. However, what differentiates agent technologies via other paradigms for software development, similar to systems that are object-oriented, systems that are dispersed, and systems of expertise, is the union of three primary characteristics (reactive nature, initiative, and social abilities) in a single thing (Rocha, Boavida-Portugal, and Gomes 2017b).

7.4 MULTI-AGENT TECHNOLOGY CLASSIFICATION

There are many alternatives for "intelligent agent" established by various researchers in an effort because of the vast range of uses, to better describe their own job, the difficulty of determining what is genuinely meaningful a "intelligent agent," as well as the tremendous momentum as this field of study has experienced in recent years. As a result, it is frequently used in the literature on specialized agents under names like robots, software agents (also known as softbots), knowbots, task bots, or user bots, personal assistants, virtual characters, etc.

Although it makes perfect sense that there may be many synonyms, they can occasionally obscure the notion itself, making it together more difficult to characterize the resource used in MAT studies. Analysing the many agent typologies suggested in the literature might help classify agents into groups with the aim to improve define discoveries from science field in which they are studied. The large number of features that have already been mentioned makes it possible to see how challenging it is to construct an agent who includes all those characteristics. It is also connected to the concept that an agent's qualities should preferably be dependent on the application type. Researchers have typed and categorized agents using a study of their properties. A typology is a division into groups of agents with similar traits.

According to Nwana (Georgakarakou and Economides 2009), there are seven different categorization dimensions for agents as seen in Figure 7.3.

Mobility: Agents can be either mobile or stationary.Mobile agents may reside permanently or temporarily on the source machine.

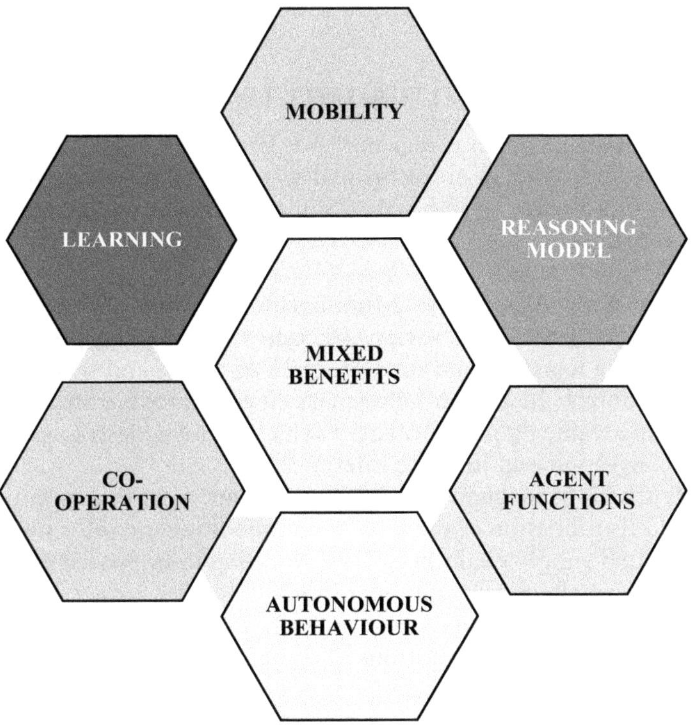

Figure 7.3 Classifications of multi-agents.

Source: Based on Geoargakarakou and Economides 2009.

Reasoning Model: Agents can exhibit either sensitivity or solely reflection, depending on the presence of certain types of symbolic reasoning.

Agent Function: Agents serve various primary roles, such as search agents seeking information or connectors facilitating man-machine communication.

Autonomous Behavior: Multi-agents are capable of exchanging information with other agents and operating independently of direct human or other agent involvement. They also possess varying degrees of control over their behavior and internal state, unaffected by external influence.

Cooperation: Agents coordinate their efforts with other agents.

Learning: Agents may possess personal, adaptive, and societal learning capacities.

Mixed Behaviors: These combine two or more distinct behavioral theories into a single agent.

7.5 APPLICATIONS OF MULTI-AGENT TECHNOLOGY

MAT, which consists of systems and methods that allow numerous agents to communicate and work in an integrated way, offers an extensive range of applications in a variety of industries. Some significant implementations of MAT are as follows:

1. **Robotics with Autonomy:** Multi-agents are used to coordinate autonomous robots in a variety of contexts, including production, warehousing logistics, and missions such as rescue and search.
2. **Traffic Control:** MAT can help smart cities improve traffic flow by combining traffic lights, automated vehicles, and pedestrians to minimize congestion and increase safety.
3. **Supply Chain Management:** Multi-agents can assist in the optimization and coordination of the flow of commodities and information in complicated supply chain networks, resulting in improved effectiveness and cost-effectiveness in operations.
4. **Agriculture:** Multi-agents are commonly used in precision agriculture to coordinate the operations of drones, automated tractors, and sensors to enhance the management of crops, water supply, and pest control.
5. **Trading and Finance:** In financial markets, MAT is utilized for algorithmic trading and analysis of the market, in which a few agents execute trade decisions almost instantaneously.
6. **The Healthcare Sector:** MASs can help with managing patient care cooperation, the allocation of resources for healthcare infrastructure, and medical equipment and patient health tracking.
7. **Smart Grids:** In smart grid management, MASs are used to streamline the flow of energy, keep an eye on the production and distribution of energy, as well as effectively address peak demand.
8. **Cobots (Collaborative Robotics):** MATs are utilized in manufacturing to build collaborative robots (cobots) that can operate safely and productively alongside human operators.
9. **Defence and Military:** MAT has been deployed to improve situational awareness as well as response capabilities in command-and-control systems manned and unmanned aerial vehicles (UAVs), especially automated military vehicles.
10. **Home Automation and IoT:** To maximize energy usage, security, and convenience, MAT can improve the administration and automation of devices in smart homes.

7.6 TOTAL QUALITY MANAGEMENT

Total quality management, also known as TQM, is a critical technique for modern firms to enhance the continuous process of decreasing and reducing manufacturing defects. It aims to enhance the customer experience, teach staff to work more effectively, and boost customer happiness. Furthermore, it promotes continuous development of internal procedures to increase the quality of a business's output, including products and services (Snongtaweeporn et al. 2020).

TQM is a theory of management founded on a series of philosophical concepts aimed at mobilizing organizational assets to maximize stakeholder satisfaction. Although firms are increasingly implementing TQM, the outcomes are not always satisfactory. Because of the disparities in the execution of TQM results, researchers and practitioners must identify the variables that underlie the achievement and failure of TQM implementation (Das, Kumar, and Kumar 2011). The term "Total" suggests that every individual inside the organization is obliged to improve operations, from invention to execution to fulfilment, and this is the primary idea underlying complete quality management. Furthermore, the term "management" suggests that this process should be a focused effort to actively monitor the quality of goods and services on a regular basis. Leadership should provide resources, people, training, and well-defined objectives. To properly comprehend the impact of TQM, upon efficiency, it is necessary to first establish the four functional pillars that TQM is intended to bring together via the implementation of a shared set of practices and methodologies as seen in Figure 7.4. Customer focus is the first functional pillar of TQM, and all quality initiatives should be driven by customer satisfaction. Therefore, TQM strives to match the goods and services it offers with customer expectations to boost satisfaction and create long-term customer loyalty.

The second pillar of TQM is one of ongoing progress or continuous improvement. This goal is most frequently achieved by utilizing a structured approach of benchmark and self-evaluation. Employee involvement or participation is the third functional pillar therefore, by allowing employees to make their own decisions and accept the responsibility of their work, the organization promotes a sense of devotion and accountability. The fourth pillar is a process-oriented approach, which comprises developing organizational procedures so that they may work as an integrated and complete system: The management system contains processes and measurements to help accomplish goals. Processes define the interrelated activities and checks that must be carried out to get the intended results (Snongtaweeporn et al. 2020).

Figure 7.4 Four pillars of TQM.

Source: Based on Snongtaweeporn et al. 2020.

7.7 ORIGINS OF TOTAL QUALITY MANAGEMENT

The foundations of TQM may be found in 1949 when the National Union of Japanese Scientists and Engineers (JUSE) established a committee of academics, engineers, and government employees with the goal of boosting Japanese production and their quality of life after World War II. The group designed a training program to teach Japanese engineers about statistical quality assurance methods, with Deming's and Juran's principles as their guiding influence. Substantial statistical instruction and the broad adoption of the Deming concept by Japanese firms came after that.

However, as quality control activities grew more commonly employed and advanced, the TQM philosophy could be applied to non-manufacturing processes such as product development, purchasing, and payment, with potential applications in service firms and nonprofits. Around 1980, American corporations began to seriously consider TQM because some US policy experts feared that Japanese manufacturing quality had met or even exceeded US standards and that Japanese productivity would soon overtake American productivity (Powell 1995).

7.8 WHAT IS THE NEED FOR TOTAL QUALITY MANAGEMENT AMONG INDUSTRIES?

The idea of "Total Quality" becomes crucial in today's age of globalization and fierce competition. Total quality requires both internal and external effectiveness in the services provided. The technique of controlling the entire quality of organisational activities to attain brilliance is known as Total Quality Management (TQM). The goal and concept of TQM is to continuously enhance quality via a process of organizational and cultural transformation, and not to assess the effectiveness of quality. It is driven by people, and the outcomes are clear in terms of better teamwork, corporate morale, and organizational environment, which lead to increased production and profitability. TQM is the core tenet for both manufacturing and customer-focused service businesses, with the goal of connecting to the customer at each phase of the business processes. Implementing quality efforts in manufacturing using TQM, Six Sigma, just-in-time (JIT), etc. has improved operational efficiency, decreased manufacturing costs, and raised enterprises' strategic competitiveness in many nations.

There is a lot of focus on quality management in the corporate world nowadays. In addition to meeting customer needs, high-quality goods and services also benefit various businesses by driving up revenue and sales and lowering operating costs. This is especially true when it comes to reducing finished goods and services defects, the number of reworks necessary to fix them, and customer rejection or product replacement. Businesses now recognize the value for quality management and intentionally work to uphold an exceptionally high level of quality for their goods and services. The level of international competitiveness has risen recently, especially in the industrial sector. Many businesses have turned to TQM in an effort to strengthen their competitiveness as a result of rising competition from overseas businesses and also because of the level of importance TQM serves in organizations over traditional management as shown in Figure 7.5. The degree to which TQM should be established alongside management performance assessment systems using measurements of manufacturing operations is a key issue in the deployment of TQM (Chenhall 1997).

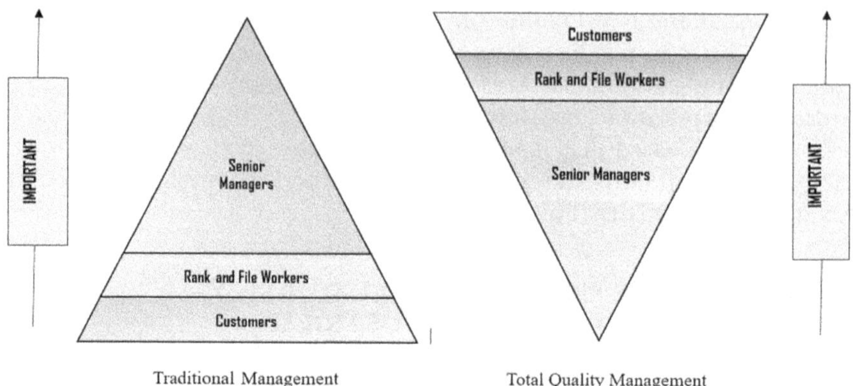

Figure 7.5 TQM: The inverse control pyramid.

Source: Adapted from Morgan and Murgatroyd 1994.

Nowadays, a lot of companies insist that their vendors hold an ISO 9001 certification, which guarantees a minimum level of quality for the products and services they buy. First and foremost, it is crucial to get the procedures correct if you want high-quality goods and services. As a result, quality management is given a lot of emphasis in companies today. Quality is no longer only a supplemental department; rather, it now plays a more integral and strategic role in enhancing procedures and quality standards inside and across multiple activities. Currently, companies without internal resources might employ outside professionals and auditors to conduct internal inspections and audits for accreditation like the International Organization for Standards, lean manufacturing, Six Sigma, etc.

7.9 THE ORGANIZATION GOALS

Based on the basic premise, an organization plays an important role in satisfying and surpassing customer demands and expectations, as well as other stakeholder expectations to a smaller extent. The interaction with the environment is not passive, but rather proactive. As a result, the surroundings must be continually surveyed. Internal procedures are becoming overburdened. The business is supposed to maintain compatibility with its surroundings. The consumer is the organization's primary stakeholder, and its quality is determined by the consumer. The notion of a customer may also be used for internal customers inside an organization. The upper management strives to ensure that all workers share values that prioritize meeting and surpassing client requirements and expectations. This, together with the idea that actuality is apparent and observable, leads to the view that both customer needs and the efficacy with which they are met can be monitored and assessed. With these assumptions, a business does not endeavour to transform the environment to meet its capabilities, i.e., it does not strive to be a market maker (Kujala and Ullrank 2004).

7.10 BENEFITS OF IMPLEMENTING TOTAL QUALITY MANAGEMENT IN ORGANIZATIONS

- Heightened competitiveness.
- The capacity to adapt to changing or developing marketplace circumstances, as well as environmental and various other regulations.
- Increased output.
- Improved market presence.
- Defects and waste are eliminated.
- Cost savings and improved management of costs.
- Increased profit.
- There has been an increase in customer focus and satisfaction.
- Retention and loyalty to customers have increased.
- Employment stability has been better.
- Employee morale has improved.
- Increased value for investors and partners.
- Processes that have been improved and are innovative.

7.11 ROLE OF MULTI-AGENT TECHNOLOGY FOR ENSURING TOTAL QUALITY MANAGEMENT IN ORGANISATIONS

In today's era, the market changes from seller-driven to customer-driven and companies play an essential part in economic growth in the dynamic period of competition, and customers now place a high priority on the products and services' quality, which supports sustainable growth. Therefore, TQM is an essential factor for saving the cost of the company and it is important for organizations to take care of it.

In simple terms, leaders must strive for a defect-free organization while reducing overall quality expenses. They must concentrate on three major types of quality expenses – internal failure, external failure, and assessment – and work on important conditions for TQM implementation success such as commitment, culture, continual development, control, and customer focus.

Hence, to overcome these issues and for an effective TTQM implementation in organizations, we use MAT, which is a computerized structure made up of several cooperating intelligent agents referred to as a system with multiple agents (MAT or "Self-Organized System") (Hu et al. 2021). Problems that are challenging or impossible for one agent or a monolithic system to address can be solved by MASs. Methodical, operational, organizational techniques, computational searching, or learning through reinforcement are examples of intelligence.

There are several benefits of MASs over conventional single-agent networks. They can perform more difficult jobs and are more scalable. Secondly, these are stronger and can withstand individual agents' failing. Therefore, organizations can use a variety of agents to more effectively solve issues related to TQM.

7.12 IDENTIFICATION OF VARIABLES OF MULTI-AGENT TECHNOLOGY INFLUENCING TOTAL QUALITY MANAGEMENT

A comprehensive literature review has been performed to identify the variables of MAT that influence the total quality of organizations. Thirty-five variables of MAT were explored from the literature and are grouped under the seven issues of TQM, as shown in Table 7.1.

7.12.1 Inventory management

Through its effects on production and delivery as well as inventory expenses, inventory management affects a company's success. You must evaluate crucial procedures and apply comprehensive quality management concepts to them to enhance inventory control. Once you are comfortable with TQM, you may utilize it to pinpoint inventory control issues and put fixes into place while increasing the effectiveness of your business operations. Inventory management may assist you in keeping a record of all your goods and figuring out the precise costs (Kaynak 2003). This demonstrates how "Inventory management" has an impact on TQM. To prevent halts, errors, reduce inventory costs in organisations, etc. in inventory operations, inventory management is divided into five MAT variables. These variables are grouped together in Table 7.1 and are addressed further below.

First, *Inventory management agents* oversee the logistics process, coordination with the goal of cost optimization given the restrictions of needed time for delivery (Dt) and delivery cost (Dc) (Giannakis and Louis 2016). The agent in the initial level is referred to as a retail agent, and those who work in the next phases are referred to as supply agents. By modifying the order's release time in response to shifts in client demand trends, the retail agent tries to achieve a target customer service level. On the other hand, each supply agent aims to synchronize the timing of its order releases with the supply of products from the upstream agent and the demand for orders from the downstream agent. (Kim, Kwon, and Kwak 2010). Thus this inventory management agent helps to manage the inventory-related issues and make TQM more effective. Second, the *Order management agent* reacts to

Table 7.1 Theoretical framework of TQM issues and variables of MAT

Sr. No.	Issues of TQM	Variables of MAT
1.	Inventory management	• Inventory management agent • Order management agent • Monitoring agent • Automated storage and retrieval system agent • Just-in-time agent
2.	World-class manufacturing	• Production planning agent • Production scheduling agent • Manufacturing disturbance agent • Manufacturing management agent • Maintenance and equipment agent
3.	Productivity	• Organizational input agent • Organizational output agent • Sales and marketing optimization agent • Customer relationship management (CRM) agent • Integration of monitoring diagnosis and prognosis agent
4.	Defects rate	• Quality control agent • Diagnosis agent • Prognosis agent • Predictive maintenance agent • Proactive maintenance agent
5.	Product design	• Virtual manufacturing agent • Simulation and analysis agent • Design collaboration agent • Human-centred design agent • Distributed digital manufacturing agent
6.	Communications within organizations	• Networking agent • Information retrieval agent • Communication routing agent • Resource allocation agent • Control system architecture agent
7.	Supplier quality management	• Supplier assessment agent • Risk assessment agent • Contract manufacturer agent • Audit and compliance agent • Big data agent

client requests by forwarding them to appropriate task agents with the help of a coordinating agent. This agent is in charge of receiving client orders, returns, cancellations, and adjustments, as well as giving contemporaneous on-demand status for consumer transactions (Giannakis and Louis 2016). Therefore, this agent helps to handle all the activities related to orders which helps to increase customer satisfaction and at the same time increases the efficiency of TQM. Third, the *Monitoring agent* is in charge of collecting and analysing pertinent data from all the cooperating partners to provide the necessary monitoring information. When an abnormal circumstance arises, the agent has the capability to sound an alert (Giannakis and Louis 2016). Thus, the monitoring agent acts as an alarm device during abnormal conditions and gives feedback to the system which helps to immediately solve the error in inventory management and helps to make TQM more economical. Fourth, *Automated storage and retrieval system agent's* primary duty is to manage the instructions given by the Digital Twin Service Systems (DTSS). The task workpiece will be transported by an Automated Guided Vehicle (AGV) after the task recipient has been determined, and an Automated Storage and Retrieval System (AS/RS) will operate in tandem with the AGV to do so (Nie et al. 2022). For that reason, this multi-agent helps to make inventory-related decisions automatically such as storage, transportation of materials, etc. Hence, it helps to increase the productivity of TQM. Fifth, *Just in time (JIT) agents* increase the efficiency of manufacturing systems by cutting costs and superfluous inventory by only ordering things as required (Rad 2006). The JIT agent supports inventory control, encourages empowerment of teams while reducing waste, and improves the flexibility and effectiveness of human resources. Thereby, as JIT agents' production process starts after customer orders, this agent helps to save inventory cost and boosts the reliability of TQM.

7.12.2 World-class manufacturing

Since the start of the Industrial Revolution, manufacturing has undergone significant change. Lean, effective, economical, and adaptable manufacturing methods are crucial for organizations in the contemporary, globally competitive era. A group of ideas known as "world-class manufacturing" establishes standards for manufacturing as well as production that other organizations may use as a guide. The idea of world-class production is linked to Japanese manufacturing as its origin. In iron and steel, electronics, and automotive industries, world-class production was implemented. Therefore, world-class manufacturing reduced inventory, easier order scheduling, and higher overall product quality, which are all benefits of smaller assembly line lot sizes (Sila and Ebrahimpour 2003). Hence, MAT is frequently used in world-class manufacturing to improve a variety of manufacturing, quality assurance, and supply chain management processes. As a result, this "world class manufacturing" factor in Table 7.1 includes five MAT variables clustered under it, which are described in more detail below.

First, *Production planning agent* –The agents within MAT manage resources, equipment, tools, and labor. Additionally, a production planning agent coordinates manufacturing activities (Singh and Sharma 2023). This agent helps the organiztions in effective production planning as the agent manages the workforce, equipment machines, etc. This agent decides all the activities such as which part is made on which machine, which person operates the specific machine and also what is the completion time of that specific task. As a result, this agent helps in smooth manufacturing process and helps to make TQm more effective. Second, *Production scheduling agent* – This specific agent is used for time management and provides equipment and raw materials to complete industrial operations in a predetermined amount of time (Singh and Sharma 2023). The agent executes the task's scheduling and replicates the shop floor manufacturing process. This agent either changes the internal agendas of the agents in the simulation mode or schedules theoretically in the planning phase (Mařík et al. 1998). Therefore, during the production process or manufacturing operation, the production scheduling agent aids in allocating, monitoring, and optimizing work and workloads, which helps to increase the speed of manufacturing operations and reduce errors, and at the same time increase the efficiency of TQM. Third, *Manufacturing disturbance agent* – This is a manufacturing disruption agent that regulates interruptions brought through low voltage, a lack of staff, delays in transport, etc. throughout the manufacturing process (Singh and Sharma 2023). This agent identifies the delays or distractions from various reasons and helps to oversee them without pauses for a continuous manufacturing process. On that account, this agent reduces disturbance and helps in smooth manufacturing by which TQM effectiveness also increases. Fourth, *Manufacturing management agent*–This MAT agent delivers input to a production planning agent (e.g., capacity availability) and conducts the manufacturing operations. The agent divides manufacturing jobs and assigns those to the relevant workshops (Giannakis and Louis 2016) which helps to complete manufacturing operations quickly and at scheduled times. This helps to boost the performance of the manufacturing operations. Fifth, *Maintenance and equipment agents* are used to avoid malfunctions and downtime; these agents keep an eye on the state of the production equipment and plan maintenance. Based on the information from sensors, they can forecast when maintenance is required (Panteleev et al., n.d.).

7.12.3 Productivity

Productivity should be considered as a measure of effectiveness as well as efficiency due to the growing global market rivalry. However, the distinction between the two is frequently missed. For instance, a business that produces in accordance with process standards may be employing its resources successfully, but only if it is producing what its consumers want. Producing well-made goods that no one would buy serves no purpose. To be lucrative, a business must evaluate productivity and value from the viewpoint of the consumer, and not from the perspective of the engineer. As a result, internal procedures can only be designed to produce goods and services that consumers consider valuable and helpful (Khan 2003). Thus, productivity is a measuring technique that assesses a business's efficiency based on the proportion of output generated to inputs utilized (Terziovski and Samson 1999). Through process optimization, improved decision-making, and job automation, MASs may be used in a variety of disciplines to boost productivity. The "productivity" element shown in Table 7.1 therefore includes five factors, which are explained below.

First, *Organizational input agent* – This agent is used to manage finances, raw materials, labour force, supply of power, and various other factors needed to start production (Singh and Sharma 2023). So this agent calculates all the activities related to starting of the production process and helps the organizations to make better decisions regarding costs and other important factors before starting the production process and helps to save time as well as money of the organization and increase its productivity. Second, *Organizational output agent* – this MAT agent includes every detail that has been collected, from raw materials to completed goods, and compiles data on the dependability, productivity, expenses, and profitability of industrial companies (Singh and Sharma 2023). Thus, this agent keeps a record of all expenses from initial to the final stage of product and helps the organization to better control it by which productivity of organization also increases. Third, *Sales and marketing optimizations agents* – They target the appropriate audience and increase conversion rates by using data and analytics to enhance sales and marketing campaigns (Calin et al. 2003). Therefore, these agents are used to forecast the sales and marketing conditions and hence used to increase TQM effectiveness and productivity of organizations. Fourth, *Customer relationship management (CRM) agents* are the agents that oversee client interactions, keep track of sales, and offer insights to raise customer satisfaction. Therefore, these agents help to build a healthy relationship with customers, and hence if relations with customers are good, productivity automatically increases (Olszak and Bartu? 2013). Fifth, *Integration of monitoring diagnosis and prognosis agent* – The monitoring, diagnosis, and prognosis integration agent serves to maintain a watch on manufacturing processes, identify faults as they occur, and take operational steps to correct them (Singh and Sharma 2023). Whenever there is any breakdown or fault detection happening during manufacturing, this agent immediately finds the causes and starts working to correct them. Therefore, this agent helps to reduce the downtime in the production process by which productivity of organization increases and further effectiveness of TQM increases.

7.12.4 Defects rate

TQM is a systematic method for enhancing product and service quality as well as customer satisfaction. TQM aims to produce these specifications with zero defects, leading to continuous improvements in output, customer satisfaction, and profits. Therefore, defect rate is the proportion of goods or services that fall short of the required requirements for quality. The number of faults in each unit, every batch, per process, or each customer may be used to calculate the defect rate (Terziovski and Samson 1999). MATs may be used to monitor, regulate, and optimize processes in the manufacturing and other sectors to lower failure rates. As a result, to lower defect rates, we will focus on five MAT variables that are shown in Table 7.1 under the "Defects Rate" component.

First, *Quality control agents* – During the production process, these agents are in charge of keeping an eye on and guaranteeing the quality of the products. To find flaws and start the necessary repair procedures, they employ sensors and data analysis (Castellini et al. 2011). Therefore, these agents help to maintain the overall required quality of products and minimize the defect rates. Second, *Diagnosis agent* – This particular agent gathers information on the effectiveness of industrial processes, diagnoses issues to minimize effort, prevents unplanned failures, and reduces maintenance costs (Singh and Sharma 2023). As it helps to reduce failures and maintenance costs, it saves the organization's cost used during preventing defects rates which further helps in reducing TQM costs of the organization. Third, *Prognosis agent* – These agents are used to determine the remaining useful life of equipment, prognosis develops a structure for fault-finding, monitoring of conditions, and programmatic prognosis as these agents predetermine the life of equipment; therefore, they help organizations to switch these tools or equipment before failure and save the breakdown time or delays. Hence, these agents help to reduce the defect rates and raise the effectiveness of TQM (Singh and Sharma 2023). Fourth, *Predictive maintenance agents* – These agents are used in industry to forecast when machinery is likely to break down, allowing for preventative maintenance to stop production interruptions that can cause faults (Panteleev et al., n.d.). Thus, they help to minimize the downtimes of machines and equipment and reduce the defect rates and hence increase the efficiency of TQM in organizations. Fifth, *Proactive maintenance agent* – When performing proactive maintenance, the agent creates an inspection to guarantee consistent and reliable output (Singh and Sharma 2023). A proactive maintenance agent aims to lower the risk of downtime and boost asset dependability. Hence, this agent also helps to reduce the defect rates associated with maintenance and boost the TQM framework.

7.12.5 Product design

A good product must foster an innovative culture, according to TQM principles, as new products are produced because of fresh ideas and new products help to generate additional revenue and net income. However, it can be used to spur the development of fresh concepts. The design of the product should ensure that it meets consumer expectations and maintains performance over the course of its useful life (Kaynak 2003). When it comes to facilitating cooperation, fostering innovation, and streamlining decision-making, MATs can be helpful. Consequently, five variables have been combined into the "Product Design" component in Table 7.1 as shown below.

First, *Virtual Manufacturing Agent* – The virtual manufacturing agent creates a manufacturing system that is computer-integrated and updates data on both the procedure and product (Singh and Sharma 2023). Virtual manufacturing agents help organizations to launch new production smoothly as it allows them to perform all the real operations virtually with the help of software on new products; therefore if there is any problem with a product or any errors happen during virtual manufacturing, organizations have the chance to immediately change its specifications according to their convenience. So, this agent helps the organization to save costs, make an accurate product design as per specifications, and to increase the effectiveness of TQM. Second, *Simulation and analysis agents* are the agents with simulation and analytic skills that can evaluate a product design's performance, structural soundness, and other attributes. They offer perceptions and suggestions for enhancing the design (Siebers and Aickelin 2011). Third, *Design collaboration agents* offer a forum for the exchange of concepts, designs, and criticism. These agents facilitate cooperation between designers, engineers, and other stakeholders. Therefore, these agents give an independent environment to share different ideas and concepts related to design aspects hence helping to make an effective product design (Liu, Tang, and Frazer 2004). Fourth, *Human-centred design agents* – These are the agents who prioritize ergonomics and user experience, making sure that the design satisfies the requirements and desires of end users. These MASs can cooperate to enable a comprehensive method of product design. Designers and engineers may speed up the design process, enhance cooperation, and ultimately produce creative and optimal products by utilizing the capabilities of these agents (Flathmann, McNeese, and Canonico 2019). Fifth, *Distributed digital manufacturing agent* helps in the manufacturing process to turn conceptual information into realities (Singh and Sharma 2023). It is a joined-up manufacturing agent that integrates digital tools with physical manufacturing execution and can be created by modelling and simulating processes to increase the quality of manufacturing facilities' decision-making while also enhancing the processes to reduce costs, speed up time to market, and create cost savings. Therefore, this helps in fast and efficient product design and it may increase effectiveness in TQM.

7.12.6 Communications within organizations

For a business to operate effectively and set up sufficient techniques, communication is crucial in TQM. Each employee in a firm needs to have efficiency and requires the capacity for transparent communication. The process of exchanging thoughts, actions messages, or information through voice, gestures, signs, or written words is known as communication. The emphasis on communication evolved from management's emphasis on human connections. Work in quality management is carried out under greater pressure compared to any other field. Due to the unique characteristics of this industry, communications in quality management differs from conventional communication. Good communication among partners is a critical component of an effective Quality Management Program. It serves as a critical link among all aspects of TQM. Communication entails the exchange of thoughts between the sender and the recipient. TQM calls for communication both between and among all employees, clients, and suppliers of the business (Huq and Stolen 1998). Therefore MATs can be useful tools for internal communication in businesses. A set of autonomous agents known as MATs collaborate to accomplish predetermined objectives. As a result, five variables have been combined in Table 7.1 in the "Communications Between Organizations" component, which is detailed below.

First, *Networking agents* refer to an internet operator-based system that allows agent-to-agent interaction to overcome challenges related to agent coordination and communication (Singh and Sharma 2023). These networking agents help to make production process flawless, as agents are communicating with each other. Therefore, if any system within the organization is not working properly, then these agents react immediately and send feedback to the system and assign that task to another agent, thus helping in better communication within organizations and enhancing the potency of TQ M in organizations. Second, *Information retrieval agents* gather information and retrieve it from a variety of sources, including databases, papers, and the internet, which is the responsibility of these agents. They offer employees pertinent data for analysis and decision-making (Luo and Xue 2010). Therefore, these multi-agents help in effective decision-making and information exchange. Third, *Communication routing agents* are those agents who oversee the distribution of letters, emails, or requests to the proper individuals or divisions inside the company. They make certain that communication is timely and effective (Zeng, Xu, and Chen 2020). Hence, these agents help to make faster communication between users within the organizations. Fourth, *Resource allocation agents* – Based on organizational requirements and resource availability, these agents oversee the distribution of resources such as meeting spaces, technology, or staff (Vizuete et al. 2022). Therefore, they maximize the use of resources available within the organizations. Fifth, *Control system architecture agent* – It is used to maintain a system that allows users to interface with the system directly. The control system architecture agent is a self-sufficient entity that functions either in online or offline mode (Singh and Sharma 2023). Therefore, this autonomous agent helps in better understanding and communication; additionally, if communication within the organization is greater, the impact of TQM is also increased as communication is also an important issue in TQM.

7.12.7 Supplier quality management

Every firm requires a scalable and adaptable supplier quality management (SQM) strategy in today's competitive and unpredictable worldwide marketplace to constantly fulfil their compliance and quality objectives by providing the highest quality products. To produce the finished item, the raw materials needed by producers must be provided by suppliers. Obstacles and incompatible materials can lead to goods recalls, lost revenue opportunities, reputational damage, and regulatory issues. Due to their importance to the manufacturing line, the vendor is also responsible for this. Raw materials must be delivered on schedule and in pristine shape. The quality of suppliers can be the criterion used to determine whether a provider meets the specified business requirements. Simply said, supplier quality is the capacity of a supplier to offer customers high-quality goods or services that meet their expectations and comply with specifications. To manage the quality of suppliers using a proactive and cooperative strategy, every firm needs an efficient SQM. SQM enables businesses to enhance the calibre of their supply chain, reduce expenses, and obtain a competitive edge (Das, Kumar, and Kumar 2011). Thus, the use of multi-agents in SQM within enterprises may be very advantageous. To make sure suppliers are meeting the standards and expectations of the company, SQM includes monitoring, evaluating, and improving supplier quality and performance. Therefore, for better SQM rates, we focus on five MAT variables that are shown in Table 7.1 under the "SQM" factor.

First, *Supplier assessment agents* are based on predetermined criteria, including product quality, on-time delivery, and adherence to contractual obligations; these agents are in charge of assessing supplier performance. Data is gathered from a variety of sources, including audits, inspections, and customer feedback (Ghadimi, Toosi, and Heavey 2018). Hence, these agents are responsible for supplier selection by rating them for different parameters, generated by organizations for selecting the best suppliers. Second, *Risk assessment agents* are the agents that examine possible hazards connected to suppliers, like monetary instability, geopolitical unrest, or supply chain interruptions. They assist in proactively detecting and reducing hazards (Giannakis and Louis 2011). Therefore, they help to smooth supplies of goods from suppliers without any risk. Third, *Contract manufacturer agents* are those MAT agents who are actually in a position of selecting dependable contractors on their behalf by the manufacturer if available capacity is insufficient to satisfy high demands (Giannakis and Louis 2016). Hence, this agent helps in better selection of vendors or suppliers for a smooth and continuous production process with low cost and quality product, therefore enhancing the effectiveness of TQM in organizations. Fourth, *Audit and compliance agents* – These are responsible for suppliers' inspection on a regular basis to make sure that quality requirements, laws, and contractual commitments are being followed. They provide audit reports and monitor remedial activities (Benanbaar, Moussaid, and Medromi 2017). Thus, they take care of all activities of suppliers on a timely basis and hence help in proper supplier selection. Fifth, a *Big data agent* is an autonomous entity or agent that oversees analysing massive data delivered by a number of providers (both public and corporate databases). It can use the semantic web to analyse and interpret data (Giannakis and Louis 2016). Big data agents can provide accurate information on organizational expenditure trends for the purpose of maintaining supplier relationships. Big data agents therefore aid in supplier selection while also fostering positive supplier relationships, which lowers TQM expenses and boosts its effectiveness.

7.13 CONCLUSION

In this chapter, we discussed how MAT ensures TQM in organizations. This study's goal is to evaluate how MAT affects overall quality control. Therefore, here we have taken seven issues from TQM, namely Inventory management, Productivity, World-class manufacturing, Defect rates, Product Design, Communication within organizations, and sSQM. For solving these problems we take 35 variables from MAT, namely Inventory management agent, Order management agent, Monitoring agent, Automated storage and retrieval system agent, Just-in-time agent, Organizational input agent, Organizational output agent, Sales and marketing optimization agent, CRM agent, Integration of monitoring diagnosis and prognosis agent, Production planning agent, Production scheduling agent, Manufacturing disturbance agent, Manufacturing management agent, Maintenance and equipment agent, Quality control agent, Diagnosis agent, Prognosis agent, Predictive maintenance agent, Proactive maintenance agent, Virtual manufacturing agent, Simulation and analysis agents, Design collaboration agents, Human-centered design agents, Distributed digital manufacturing agent, Networking agents, Information retrieval agents, Communication routing agents, Resource allocation agents, Control system architecture, Contract manufacturer agent, and Big data agent.

Therefore, organizations may focus on the MAT variables listed above in TQM activities. It may increase effectiveness in TQM and ensure satisfactory outcomes. To ensure the TQM program by these MAT variables, the management should convey to the workforce a vision that elaborates on the aim of the transformation. Management should lead the implementation process and include the human resources division in the transformation process, depriving the initiative of its most probable backer. Thus, MAT may be used to ensure TQM in organizations to operate businesses effectively and efficiently, respond to consumer requests, please customers, and achieve a competitive edge in today's global market.

REFERENCES

Benanbaar, N., Moussaid, L., & Medromi, H. (2017). A Multi-Agent Systems Contribution for Audit and Change Management. *International Journal of Advanced Engineering Research and Science*, 4(8), 80–85. https://doi.org/10.22161/ijaers.4.8.14

Calin Sandru, V. N. D. P. (2003). A multi-agent approach to a sales optimization applications. *14th International Conference on Control Systems and Computer Science (CSIS-14)*.

Castellini, P., Cristalli, C., Foehr, M., Leitao, P., Paone, N., Schjolberg, I., Tjonnas, J., Turrin, C., & Wagner, T. (2011). Towards the integration of process and quality control using multi-agent technology. *IECON 2011 - 37th Annual Conference of the IEEE Industrial Electronics Society*, 421–426. https://doi.org/10.1109/IECON.2011.6119347

Chenhall, R. H. (1997). Reliance on manufacturing performance measures , total quality management and organizational performance. In *Management Accounting Research* (Vol. 8).

Das, A., Kumar, V., & Kumar, U. (2011). The role of leadership competencies for implementing TQM: An empirical study in Thai manufacturing industry. *International Journal of Quality and Reliability Management*, 28(2), 195–219. https://doi.org/10.1108/02656711111101755

Flathmann, C., McNeese, N., & Canonico, L. B. (2019). Using Human-Agent Teams to Purposefully Design Multi-Agent Systems. *Proceedings of the Human Factors and Ergonomics Society Annual Meeting*, 63(1), 1425–1429. https://doi.org/10.1177/1071181319631238

Franklin, S., & Graesser, A. (1997). Is It an agent, or just a program?: A taxonomy for autonomous agents. In *Stan Franklin and Art Graesser* (pp. 21–35). https://doi.org/10.1007/BFb0013570

Georgakarakou, C. E., & Economides, A. A. (2009). Software Agent Technology. In *Software Applications* (pp. 128–151). IGI Global. https://doi.org/10.4018/978-1-60566-060-8.ch014

Ghadimi, P., Ghassemi Toosi, F., & Heavey, C. (2018). A multi-agent systems approach for sustainable supplier selection and order allocation in a partnership supply chain. *European Journal of Operational Research*, 269(1), 286–301. https://doi.org/10.1016/j.ejor.2017.07.014

Giannakis, M., & Louis, M. (2011). A multi-agent based framework for supply chain risk management. *Journal of Purchasing and Supply Management*, 17(1), 23–31. https://doi.org/10.1016/j.pursup.2010.05.001

Giannakis, M., & Louis, M. (2016). A multi-agent based system with big data processing for enhanced supply chain agility. *Journal of Enterprise Information Management, 29*(5), 706–727. https://doi.org/10.1108/JEIM-06-2015-0050

Hu, J., Bhowmick, P., Jang, I., Arvin, F., & Lanzon, A. (2021). A Decentralized Cluster Formation Containment Framework for Multirobot Systems. *IEEE Transactions on Robotics, 37*(6), 1936–1955. https://doi.org/10.1109/TRO.2021.3071615

Huq, Z., & Stolen, J. D. (1998). Total quality management contrasts in manufacturing and service industries. *International Journal of Quality & Reliability Management, 15*(2), 138–161. https://doi.org/10.1108/02656719810204757

Kaynak, H. (2003). The relationship between total quality management practices and their effects on firm performance. *Journal of Operations Management, 21*(4), 405–435. https://doi.org/10.1016/S0272-6963(03)00004-4

Khan, J. H. (2003). Impact of total quality management on productivity. *TQM Magazine, 15*(6), 374–380. https://doi.org/10.1108/09544780310502705

Kim, C. O., Kwon, I. H., & Kwak, C. (2010). Multi-agent based distributed inventory control model. *Expert Systems with Applications, 37*(7), 5186–5191. https://doi.org/10.1016/j.eswa.2009.12.073

Kujala, J., & Ullrank, P. (2004). Total quality management as a cultural phenomenon. *Quality Management Journal, 11*(4), 43–55. https://doi.org/10.1080/10686967.2004.11919132

Liu, H., Tang, M., & Frazer, J. H. (2004). Supporting dynamic management in a multi-agent collaborative design system. *Advances in Engineering Software, 35*(8–9), 493–502. https://doi.org/10.1016/j.advengsoft.2004.06.007

Luo, J., & Xue, X. (2010). Research on information retrieval system based on Semantic Web and Multi-Agent. *Proceedings - 2010 International Conference on Intelligent Computing and Cognitive Informatics, ICICCI 2010*, 207–209. https://doi.org/10.1109/ICICCI.2010.35

Mařík, V., Pěchouček, M., Štěpánková, O., & Lažanský, J. (1998). Application of the Multi-Agent Approach in Production Planning and Modelling. *IFAC Proceedings Volumes, 31*(31), 11–16. https://doi.org/10.1016/S1474-6670(17)40997-9

Nie, Q., Tang, D., Zhu, H., & Sun, H. (2022). A multi-agent and internet of things framework of digital twin for optimized manufacturing control. *International Journal of Computer Integrated Manufacturing, 35*(10–11), 1205–1226. https://doi.org/10.1080/0951192X.2021.2004619

Olszak, C. M., & Bartuś, T. (2013). Multi-agent framework for social customer relationship management systems. *Issues in Informing Science and Information Technology, 10*, 367–387. https://doi.org/10.28945/1817

Panteleev, V. V, Kamaev, V. A., Kizim, A. V, & Matokhina, A. V. (n.d.). *Using of Multi-Agent System to Model a Process of Maintenance Service and Repair of Equipment of a Service Company.*

Powell, T. C. (1995). TOTAL QUALITY MANAGEMENT AS COMPETITIVE ADVANTAGE: A REVIEW AND EMPIRICAL STUDY. In *Strategic Management Journal* (Vol. 16).

Pulikottil, T., Estrada-Jimenez, L. A., Rehman, H. U., Barata, J., Nikghadam-Hojjati, S., & Zarzycki, L. (2021). Multi-agent based manufacturing: Current trends and challenges. *IEEE International Conference on Emerging Technologies and Factory Automation, ETFA, 2021-September.* https://doi.org/10.1109/ETFA45728.2021.9613555

Rad, A. M. M. (2006). The impact of organizational culture on the successful implementation of total quality management. *TQM Magazine, 18*(6), 606–625. https://doi.org/10.1108/09544780610707101

Reis, L. P., Lau, N., & Oliveira, E. C. (2001). Situation based strategic positioning for coordinating a team of homogeneous agents. *Lecture Notes in Computer Science (Including Subseries Lecture Notes in Artificial Intelligence and Lecture Notes in Bioinformatics), 2103,* 175–197. https://doi.org/10.1007/3-540-44568-4_11

Rocha, J., Boavida-Portugal, I., & Gomes, E. (2017a). Introductory Chapter: Multi-Agent Systems. In *Multi-agent Systems.* InTech. https://doi.org/10.5772/intechopen.70241

Rocha, J., Boavida-Portugal, I., & Gomes, E. (2017b). Introductory Chapter: Multi-Agent Systems. In *Multi-agent Systems.* InTech. https://doi.org/10.5772/intechopen.70241

Siebers, P.-O., & Aickelin, U. (2011). Introduction to Multi-Agent Simulation. In *Encyclopedia of Decision Making and Decision Support Technologies.* IGI Global. https://doi.org/10.4018/9781599048437.ch062

Sila, I., & Ebrahimpour, M. (2003). Examination and comparison of the critical factors of total quality management (TQM) across countries. *International Journal of Production Research, 41*(2), 235–268. https://doi.org/10.1080/0020754021000022212

Singh, V., & Sharma, S. K. (2023). Critical factors of multi-agent technology influencing manufacturing organizations: an AHP and DEMATEL-oriented analysis. *International Journal of Computer Integrated Manufacturing.* https://doi.org/10.1080/0951192X.2023.2209857

Snongtaweeporn, T., Siribensanont, C., Kongsong, W., & Channuwong, S. (2020). Total quality management in modern organizations by using participation and teamwork. *Journal of Arts Management Vol.4 No.3 September–December 2020.*

Terziovski, M. Â., & Samson, D. (1999). The link between total quality management practice and organisational performance. In *International Journal of Quality & Reliability Management* (Vol. 16, Issue 3). # MCB University Press.

Vizuete, R., De Galland, C. M., Hendrickx, J. M., Frasca, P., & Panteley, E. (2022). Resource allocation in open multi-agent systems: an online optimization analysis. *Proceedings of the IEEE Conference on Decision and Control, 2022-December,* 5185–5191. https://doi.org/10.1109/CDC51059.2022.9993038

Wooldridge, M., & Jennings, N. R. (1995). Intelligent agents: theory and practice. *The Knowledge Engineering Review, 10*(2), 115–152. https://doi.org/10.1017/S0269888900008122

Zeng, S., Xu, X., & Chen, Y. (2020). Multi-Agent Reinforcement Learning for Adaptive Routing: A Hybrid Method using Eligibility Traces. *2020 IEEE 16th International Conference on Control & Automation (ICCA),* 1332–1339. https://doi.org/10.1109/ICCA51439.2020.9264518

Enhancing human–machine collaboration for value creation in automotive manufacturing in Industry 5.0

Priya V., Vipin C., Zamal Mohamed Zubair, and Pranav S.

8.1 INTRODUCTION

In the tapestry of technological evolution, the industrial landscape has undergone transformative shifts, each epoch marked by distinct paradigms that have shaped the way we produce, manufacture, and innovate. From the mechanization of Industry 1.0 to the automation of Industry 4.0, the journey has been nothing short of revolutionary. Today, we stand at the precipice of a new era – Industry 5.0 – where the fusion of human ingenuity and machine capabilities takes center stage (Yang, Liu, Liu, and Morgan 2022). The evolution of industrial paradigms has been a compelling narrative, illustrating our relentless pursuit of efficiency and progress. Industry 1.0 saw the mechanization of production through water and steam power, while Industry 2.0 introduced mass production with the advent of electricity. The digital age dawned with Industry 3.0, bringing computerization and automation to the forefront, paving the way for the interconnected world we inhabit today. Industry 4.0, characterized by the rise of the Internet of Things (IoT) and artificial intelligence (AI), further propelled us into a realm where machines communicated seamlessly, creating a digital tapestry of interconnected processes. As we transition to Industry 5.0, the narrative shifts from mere automation to a profound synergy between humans and machines. This paradigm is not just about the next iteration of technology but represents a fundamental shift in our approach to industrial processes. Industry 5.0 envisions a future where human workers and intelligent machines collaborate in harmony, each leveraging their unique strengths to create a more agile, adaptable, and sustainable industrial ecosystem (Łapczyńska 2023). The essential nature of human–machine collaboration in this era cannot be overstated. Unlike its predecessors, Industry 5.0 is not about replacing human labor with machines but rather enhancing human capabilities through intelligent automation. This symbiosis acknowledges that while machines excel at repetitive tasks and data processing, humans bring creativity, critical thinking, and emotional intelligence to the table (Mourtzis, Angelopoulos, and Panopoulos 2023). It is a realization that the true potential lies in the

DOI: 10.1201/9781032677040-8

convergence of human and machine intelligence, creating a workforce that is not just efficient but also empathetic, innovative, and resilient. In this era of unprecedented collaboration, the research objective is to delve into the intricacies of Industry 5.0, understanding its implications on the workforce, businesses, and society at large. We aim to explore the nuances of human–machine interaction, examining the challenges and opportunities that arise as we navigate this uncharted territory. The scope of this research extends beyond the technical aspects of Industry 5.0 to encompass the socioeconomic, ethical, and cultural dimensions, recognizing that the impact of this paradigm shift transcends the boundaries of technology. The structure of this paper is designed to unfold the narrative (Maddikunta, Pham, Prabadevi, Deepa, Dev, Gadekallu, ... and Liyanage 2022). We will begin by delving into the historical trajectory of industrial paradigms, setting the stage for the emergence of Industry 5.0. From there, we will navigate the intricate dance between humans and machines, exploring how this collaboration is reshaping industries and redefining the very fabric of work. The discussion will extend to the broader implications on society, addressing concerns and celebrating the potential benefits that arise from this transformative era. Finally, we will conclude the study by envisioning the roadmap ahead, contemplating the ethical considerations and societal frameworks needed to harness the full potential of Industry 5.0 while ensuring a future that is not only technologically advanced but also human-centric and sustainable (Mekkunnel 2019).

8.2 HUMAN–MACHINE COLLABORATION FRAMEWORK FOR AUTOMOTIVE MANUFACTURING

In the ever-evolving landscape of automotive manufacturing, the intersection of human expertise and machine precision stands as a promising frontier. As we strive to redefine efficiency, quality, and innovation in this industry, the need for a comprehensive Human–Machine Collaboration Framework becomes increasingly evident (Longo, Padovano, and Umbrello 2020). The proposed framework aims not only to streamline processes but also to create a harmonious synergy where human creativity and technological prowess amplify each other. In the following discussion, we will delve into the aims of this framework, dissect its key components, and explore strategies and principles vital for ensuring effective collaboration in the dynamic realm of automotive manufacturing.

8.2.1 Objectives of the proposed framework

The overarching goal of our Human–Machine Collaboration Framework is to revolutionize automotive manufacturing by fostering a collaborative ecosystem where humans and machines seamlessly integrate their strengths. We seek to achieve the following key objectives:

1. Enhanced Efficiency and Precision: The framework aims to optimize manufacturing processes by harnessing the precision of machines for tasks that require accuracy while allowing human workers to focus on complex, creative, and decision-intensive aspects of the production chain.
2. Innovation Catalyst: By promoting collaboration, the framework intends to create an environment that stimulates innovation. Human creativity, critical thinking, and problem-solving skills, when combined with machine learning (ML) and automation, can lead to breakthroughs in product design, process optimization, and overall industry advancement.
3. Adaptive Resilience: Recognizing the dynamic nature of the automotive industry, the framework seeks to instill adaptability and resilience. Humans and machines working in tandem can respond more effectively to changing demands, market trends, and unforeseen challenges.

8.2.2 Key components of the framework

1. **Intelligent Automation Systems:** At the heart of the framework lies the deployment of intelligent automation systems. These systems, integrating robotics and AI, handle repetitive and data-driven tasks, allowing human workers to focus on tasks that require creativity, intuition, and complex decision-making.
2. **Collaborative Workstations:** Physical spaces are designed to facilitate collaboration between humans and machines. These workstations are equipped with advanced human–machine interfaces, ensuring a seamless exchange of information and feedback between the two entities.
3. **Skill Development Initiatives:** To ensure the workforce is ready for this paradigm shift, the framework includes robust skill development programs. These initiatives aim to upskill human workers, empowering them with the knowledge and expertise required to collaborate effectively with advanced technologies.
4. **Data Analytics and Decision Support Systems:** Leveraging the power of big data, the framework incorporates advanced analytics and decision support systems. This enables real-time monitoring of manufacturing processes, data-driven insights, and informed decision-making for both humans and machines.

8.2.3 Strategies and principles for effective collaboration

1. **Clear Communication Channels:** Establishing clear and open communication channels is fundamental. This involves designing interfaces that facilitate easy interaction between humans and machines, ensuring that information is conveyed comprehensively and intuitively.
2. **Human-Centric Design:** The framework prioritizes human-centric design principles, acknowledging the importance of user experience. Interfaces, controls, and collaboration tools are designed with the user in mind to minimize cognitive load and enhance usability.
3. **Continuous Learning Culture:** Embracing a culture of continuous learning is crucial for success. Both humans and machines should be adaptable, with ongoing training programs and updates to keep pace with evolving technologies and industry requirements.
4. **Ethical Considerations and Transparency:** As collaboration intensifies, ethical considerations become paramount. The framework incorporates transparent decision-making processes and emphasizes the ethical use of technology to ensure that human–machine collaboration aligns with societal values.

8.3 EMERGING TECHNOLOGIES FOR ENHANCED HUMAN–MACHINE COLLABORATION

8.3.1 Overview of emerging technologies for enhanced human–machine collaboration

In our quest for Industry 5.0, a plethora of emerging technologies are reshaping the landscape of human–machine collaboration, ushering in an era where synergy between the two becomes not just a possibility but a necessity. Let's explore the intricacies of each technological area and understand how they contribute to the enhancement of collaboration.

1. **AI:** AI, often deemed the cornerstone of Industry 5.0, is revolutionizing the way machines and humans work together. ML algorithms enable systems to learn and adapt, providing unprecedented levels of efficiency. From predictive analytics to natural language processing, AI empowers machines to understand, interpret, and respond to human inputs in real time. This fosters a collaborative environment where AI becomes a cognitive partner, augmenting human decision-making processes rather than replacing them.

    Relevance to Collaboration:

    AI facilitates data-driven decision-making, automates routine tasks, and augments human capabilities, allowing for more focused and creative collaboration. In sectors like healthcare, AI assists in diagnosis and treatment planning, enabling healthcare professionals to make more informed decisions.

2. **ML:** Working in tandem with AI, ML algorithms empower systems to improve their performance over time without explicit programming. In human–machine collaboration, ML enables machines to recognize patterns, make predictions, and continuously evolve based on data inputs.

    Relevance to Collaboration:

    ML enhances collaboration by automating complex tasks, such as predictive maintenance in manufacturing. Machines equipped with ML algorithms can predict equipment failures, allowing humans to proactively address issues, thus reducing downtime and optimizing operational efficiency.

3. **Advanced Robotics:** Beyond traditional automation, advanced robotics combines AI and sophisticated mechanics, enabling machines to perform intricate tasks with precision. These robots, ranging from humanoid to specialized industrial machines, are designed to collaborate seamlessly with human workers.

    Relevance to Collaboration:

    In manufacturing, advanced robotics works alongside human workers to handle tasks that require precision and strength. This collaborative approach optimizes production processes, with robots taking on repetitive or dangerous tasks, allowing humans to focus on tasks that require creativity and problem-solving.

4. **Collaborative Robots (Cobots):** Cobots represent a specific subset of robotics designed explicitly for collaboration with humans. Unlike traditional robots confined to safety cages, cobots are equipped with sensors and intelligence to work alongside humans without posing a threat.

    Relevance to Collaboration:

    Cobots find applications in various industries, such as assembly lines and healthcare. In manufacturing, they collaborate with human workers to assemble products, enhancing efficiency and ensuring a safer working environment.

5. **Simulation:** Simulation technologies provide a virtual environment for testing and refining processes before implementation. This not only reduces the risk of errors but also enhances collaboration by allowing humans and machines to fine-tune their interactions in a controlled digital space.

Relevance to Collaboration:

Industries such as aerospace use simulation for training pilots and testing new aircraft designs. In a collaborative context, simulation allows humans and machines to refine their coordination and communication strategies.

6. **Virtual Reality (VR):** VR immerses users in a simulated environment, providing a new dimension to human–machine collaboration. It enables remote collaboration, allowing individuals to interact with each other and virtual elements as if they were physically present.

   Relevance to Collaboration:

   In fields like architecture and design, VR facilitates collaborative design sessions where team members, regardless of geographical location, can interact with and modify virtual prototypes in real time, fostering a more inclusive and dynamic collaborative process.

7. **Digital Twins:** A digital twin is a virtual replica of a physical object, process, or system. It allows real-time monitoring, analysis, and optimization, offering a dynamic representation of the physical counterpart.

   Relevance to Collaboration:

   In manufacturing, digital twins enable humans and machines to collaborate by providing a shared understanding of the production processes. This shared digital representation enhances communication, enabling quick adjustments and improvements based on real-time data.

8. **Big Data Analytics:** The massive volumes of data generated in the digital age are harnessed through big data analytics. This involves processing and analyzing large datasets to extract valuable insights, driving informed decision-making.

   Relevance to Collaboration:

   Big data analytics plays a crucial role in optimizing collaboration by providing actionable insights. In logistics, for instance, analytics can predict demand patterns, facilitating efficient collaboration between suppliers and manufacturers to meet customer needs.

9. **Internet of Things (IoT):** IoT connects physical devices and objects to the Internet, enabling them to collect and exchange data. This interconnectedness creates a network where devices collaborate in real time to enhance efficiency and responsiveness.

   Relevance to Collaboration:

   In smart cities, IoT fosters collaboration between infrastructure elements. For example, traffic lights communicate with vehicles, optimizing traffic flow. This interconnectedness enhances overall system performance through collaborative decision-making between machines.

In this era of Industry 5.0, these emerging technologies converge to create a collaborative ecosystem where the strengths of humans and machines complement each other. This symbiosis is not just about technological advancement but about creating a harmonious and efficient coexistence that transcends traditional boundaries, shaping a future where collaboration is the driving force behind innovation and progress (Adel 2022). The key technologies enabling this human–machine collaboration are summarized in Table 8.1, along with details on the algorithms, models, and use cases. Furthermore, Figure 8.1 illustrates the progression of industrial revolutions leading up to Industry 5.0, which is characterized by this concept of advanced human–machine collaboration.

Table 8.1 Enhanced technologies and algorithm details used in deployment

Technology	Algorithms/Models Used	Details	Use Cases
AI	Deep neural networks, convolutional neural networks (CNN), recurrent neural networks (RNN), graph neural networks (GNN)	CNNs for computer vision, RNNs for sequence data, GNNs for network topology data. Various neural network architectures and training techniques used.	Automated quality inspection, customized product design, human–robot dynamic planning
ML	**Supervised:** Regression – linear regression, logistic regression, support vector machine (SVM), decision trees, random forest. **Classification:** SVM, KNN (K-nearest neighbour) KNN, random forest, Ada Boost. **Unsupervised:** Clustering, K-means, hierarchical clustering, density-based spatial clustering of applications with noise (DBSCAN), anomaly detection algorithms, optimization algorithms.	Combinations of feature extraction, feature selection, cross-validation, hyper parameter tuning used. Accuracy metrics important for performance. Both classical ML and deep learning techniques applied.	Predictive maintenance, inventory optimization, electricity demand forecasting

(continued)

Table 8.1 (Cont.)

Technology	Algorithms/Models Used	Details	Use Cases
Advanced Robotics	Imitation learning, inverse reinforcement learning to replicate human demonstrations. Reinforcement learning using policy and value networks as function approximators trained with reward signals.	Techniques to map complex human tasks into appropriate reward functions and state/action spaces for robots. Sim2Real transfer for robust models. Safety constraints important.	Human–robot teams for manufacturing with dynamic task allocation leveraging strengths
Collaborative Robots	Computer vision uses 3D sensors, algorithms like random sample consensus (RANSAC) for robust model fitting. High-frequency motion emergency stops.	Key focus areas are precision, flexibility, and human safety. Must avoid single points of failure. System design informed through risk assessment.	Collaborative assembly where human handles complexity while robot does repetitive tasks

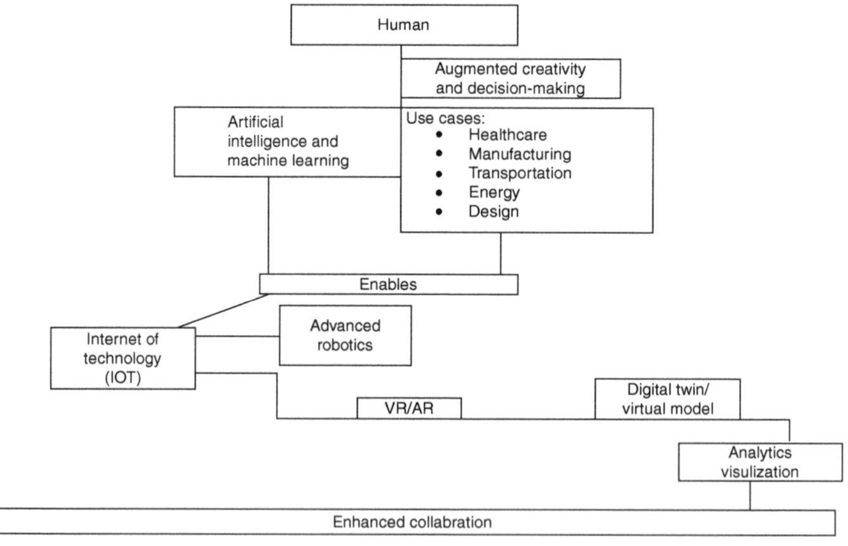

Figure 8.1 Key technologies enabling human–machine collaboration in Industry 5.0.
Source: Own study.

8.4 UNVEILING THE SYNERGY OF HUMAN–MACHINE COLLABORATION

In the intricate dance between humans and machines within Industry 5.0, the true measure of success lies in the tangible value it brings to various facets of our lives. This section unveils a tapestry of practical examples showcasing how human–machine collaboration creates value, not just in terms of efficiency and productivity, but also in fostering innovation, enhancing safety, and thereby ultimately reshaping the landscape of industries (Pizoń and Gola 2023; Zhang, Wang, Zhou, Chang, Ma, Jing, …, and Zhao 2023).

1. **Manufacturing Precision with Collaborative Robots:** In the realm of manufacturing, collaborative robots, or cobots, exemplify the seamless integration of human and machine capabilities. Take the example of a production line where cobots work alongside human counterparts to assemble intricate components. These robots excel in precision and repetition, ensuring consistent quality, while human workers contribute their problem-solving skills and adaptability. The result is a harmonious assembly process that combines the efficiency of machines with the nuanced decision-making abilities of humans.

2. **Healthcare Augmentation through AI Diagnostics:** In the healthcare sector, the collaboration between human medical professionals and AI is transforming diagnostics. AI algorithms analyze vast datasets to detect patterns and anomalies, providing physicians with valuable insights for accurate diagnosis. This synergy expedites the diagnostic process, allowing healthcare providers to focus on personalized patient care. The combination of AI's analytical prowess and human empathy leads to improved patient outcomes and a more efficient healthcare system. Specifically, convolutional neural networks (CNNs) have been widely adopted for healthcare applications such as disease modeling, medical image segmentation, and multimodal data fusion. These deep learning architectures enable intelligent analysis of imaging data and sensor inputs, augmenting human expert decision-making in numerous clinical workflows.

3. **Enhancing Creativity in Design Studios:** The creative realms are not exempt from the transformative influence of Industry 5.0. Design studios, for instance, leverage generative design algorithms that work in tandem with human designers. These algorithms explore countless design possibilities based on specified parameters, freeing designers from routine tasks and enabling them to channel their creativity into refining and selecting the most innovative solutions. The collaborative interplay between human intuition and machine-generated possibilities results in groundbreaking designs that push the boundaries of conventional thinking.

4. **Logistics Optimization with Smart Warehousing:** In the logistics sector, the integration of human expertise and automated systems has led to the development of smart warehouses. Autonomous robots navigate the warehouse floors, retrieving and organizing inventory, while human workers manage complex logistical decisions and strategic planning. This collaborative approach ensures efficient order fulfillment, minimizes errors, and optimizes inventory management. The real-time decision-making abilities of humans complement the precision and speed of automated systems, creating a resilient and responsive supply chain (Adel 2023).

5. **Empowering Financial Decision-Making:** In the financial industry, the collaboration between human financial analysts and robo-advisors exemplifies the value creation in decision-making processes. Robo-advisors analyze market trends, assess risk, and recommend investment strategies, while human analysts provide the nuanced understanding of market dynamics, geopolitical influences, and client preferences. This symbiotic relationship results in more informed and well-rounded investment decisions, combining the analytical capabilities of machines with the strategic insights of human experts.

6. **Education Reinvented through Personalized Learning:** The education sector embraces Industry 5.0 by incorporating personalized learning platforms. AI algorithms analyze student performance data, identifying individual learning styles and areas for improvement. Teachers, armed with this information, can tailor their instructional approaches to meet the unique needs of each student. The collaboration between AI-driven insights and human educators fosters a more engaging and effective learning environment, nurturing the diverse talents and learning preferences of students.

7. **Safety Augmentation in Manufacturing Environments:** In manufacturing environments, the integration of wearable technologies and AI-powered monitoring systems enhances workplace safety. Workers wear smart helmets equipped with sensors that detect environmental hazards and provide real-time feedback on potential risks. AI algorithms process this data, predicting and preventing accidents. Human workers, armed with this augmented awareness, can proactively address safety concerns and contribute to the continuous improvement of safety protocols (Sindhwani, Afridi, Kumar, Banaitis, Luthra, and Singh 2022).

8. **Agricultural Precision through Smart Farming:** The agricultural sector embraces the benefits of human–machine collaboration through smart farming practices. Drones equipped with sensors monitor crop health, while AI algorithms analyze data to identify areas requiring

attention. Human farmers, informed by these insights, can make strategic decisions on irrigation, pest control, and harvesting schedules. This collaborative approach maximizes crop yields, minimizes resource usage, and contributes to sustainable agricultural practices.

9. **Customer Service Excellence with AI Chatbots:** In the realm of customer service, AI chatbots exemplify how human–machine collaboration enhances efficiency. Chatbots handle routine queries and transactions, freeing human customer service representatives to focus on complex issues requiring emotional intelligence and empathy. This collaborative approach ensures faster response times, 24/7 availability, and improved customer satisfaction, showcasing the value of combining machine efficiency with human warmth.

10. **Innovative Product Development:** The product development cycle undergoes a metamorphosis with Industry 5.0, where human creativity and machine-driven simulations intersect. Engineers and designers collaborate with AI-powered simulations to explore design iterations, analyze performance under various conditions, and predict potential challenges. This collaborative innovation process accelerates product development, reduces costs, and brings cutting-edge products to market faster (Peruzzini, Prati, and Pelicciari 2023).

8.5 LIMITATIONS IN THE AGE OF INDUSTRY 5.0: NAVIGATING POTENTIALS AND PITFALLS

In the tapestry of progress woven by Industry 5.0, the potentials seem boundless, promising a future where humans and machines collaborate seamlessly for unprecedented innovation and efficiency. However, within this landscape of promise, there lie challenges and limitations that warrant careful consideration. This section aims to dissect the nuanced aspects of Industry 5.0, exploring how the dreams of collaboration might fall short and delving into the ethical considerations that accompany the adoption of transformative technologies.

8.5.1 The potentials of Industry 5.0: A glimpse into the future

8.5.1.1 Enhanced productivity and efficiency

At the heart of Industry 5.0 lies the promise of unparalleled productivity and efficiency. The collaboration between humans and machines, each leveraging their unique strengths, holds the potential to streamline processes, reduce errors, and optimize resource utilization.

8.5.1.2 Innovation and creativity amplified

In the collaborative dance between human intellect and machine precision, innovation takes center stage. Industry 5.0 envisions a future where creative solutions to complex problems emerge from the synergy of human intuition and ML. It is not just about automating existing processes but reimagining them, pushing the boundaries.

8.5.1.3 Adaptive and resilient systems

The adaptive nature of Industry 5.0 systems is a testament to their resilience. Machines equipped with AI can learn and adapt to changing circumstances, creating a dynamic ecosystem that can weather unforeseen challenges. This adaptability extends beyond technical aspects, encompassing the ability of human workers to upskill and adapt to evolving roles.

8.6 FUTURE SCOPES

As we navigate the exciting terrain of Industry 5.0 and the evolving landscape of human–machine collaboration, it becomes imperative to identify potential spaces for future research. These avenues of exploration emerge from the limitations and gaps identified in existing literature, offering a roadmap for scholars and practitioners to delve deeper into the intricacies of this transformative paradigm (Ordieres-Meré, Gutierrez, and Villalba-Díez 2023).

8.6.1 Enhancing ethical frameworks in human–machine collaboration

One of the key limitations observed in the current discourse is the need for a robust ethical framework governing human–machine collaboration. Future research could delve into developing comprehensive guidelines that address issues such as data privacy, algorithmic bias, and the ethical implications of AI-driven decision-making. This research direction aims to ensure that as Industry 5.0 advances, it does so within a framework that prioritizes fairness, transparency, and accountability.

8.6.2 Reskilling and upskilling for the workforce of the future

The transition to Industry 5.0 raises pertinent questions about the preparedness of the global workforce. Future research should explore innovative approaches to reskilling and upskilling programs, considering the dynamic nature of job roles in a landscape where humans and machines collaborate closely. This research direction aims to bridge the gap between traditional education systems and the skills demanded by the evolving industrial paradigm.

8.6.3 Cultural impacts of human–machine collaboration

The cultural dimension of Industry 5.0 is a relatively underexplored area. Future research could investigate how different cultures perceive and adapt to the increasing integration of intelligent machines in the workplace. Understanding cultural nuances can inform the development of strategies that foster inclusive collaboration and mitigate potential resistance to technological change.

Table 8.2 Research areas and implication

Research Area	Objectives	Methods	Implications
Enhancing Ethical Frameworks	Develop guidelines for data privacy, algorithmic fairness, responsible AI	Surveys of current practices, risk–benefit analyses, focus groups with stakeholders	Ensure human values are upheld as Industry 5.0 progresses
Reskilling and Upskilling	Create and evaluate training programs for dynamic new roles	Training needs assessments, program outcome studies	Bridge skills gap by aligning education with evolving jobs
Cultural Impacts	Understand cultural perceptions and adoption of human–AI collaboration	Multi-region qualitative studies, technology acceptance models	Strategies for inclusive collaboration across diverse cultures
Human-Centric AI Design	Incorporate human factors like user experience (UX), cognitive alignment into AI system design	User testing, participatory design, experimental psychology methods	AI systems personalized for end users; reduced complexity
Responsible Innovation	Develop frameworks for balancing productivity and societal good	Analysis of regulations, focus groups with public stakeholders	Ensure Industry 5.0 innovations benefit society holistically
Mental Health Considerations	Study psychological effects of extensive human–AI interactions	Longitudinal observational studies, clinician surveys	Support worker well-being as automation increases

Source: Own study.

8.6.4 Human-centric design in AI systems

The design of AI systems often neglects the human element. Future research should explore suitable methodologies for incorporating human-centric design principles into the development of intelligent systems. This includes considerations for user experience, interface design, and the integration of AI technologies that align with human cognitive processes, preferences, and emotional responses.

Table 8.2 summarizes key avenues for future research needed to ensure responsible and ethical progression of human–machine collaboration paradigms across various facets of Industry 5.0.

8.7 CONCLUSION

In concluding this exploration of Industry 5.0 and the collaborative dance between humans and machines, we find ourselves at the intersection of innovation, challenge, and boundless potential. Having traced the historical evolution of industrial paradigms, Industry 5.0 emerges as a pivotal shift, emphasizing a profound symbiosis between human creativity and machine efficiency (Johri, Singh, Sharma, and Rastogi 2021). Key considerations include the imperative development of ethical frameworks, the necessity of reskilling for a dynamic workforce, and the acknowledgment of cultural impacts on collaboration. As we envision the future, the prospects are both exhilarating and challenging: augmented creativity in the arts, a healthcare revolution through human–AI collaboration, inclusive economic growth, humanitarian applications of AI, and sustainable urban development (Nixdorf, Ansari, and Schlund 2022). This era calls for a holistic approach, where technological progress aligns with ethical principles, fosters inclusivity, and propels humanity toward a future where the collaborative synergy between humans and machines is not just a technological marvel but a purpose-driven force for positive societal transformation.

REFERENCES

Adel, A. (2022). Future of industry 5.0 in society: Human-centric solutions, challenges and prospective research areas. *Journal of Cloud Computing*, *11*(1), 1–15.

Adel, A. (2023). Unlocking the Future: Fostering Human–Machine Collaboration and Driving Intelligent Automation through Industry 5.0 in Smart Cities. *Smart Cities*, *6*(5), 2742–2782.

Johri, P., Singh, J. N., Sharma, A., & Rastogi, D. (2021, December). Sustainability of coexistence of humans and machines: An evolution of industry 5.0 from industry 4.0. In 2021 10th International Conference on System Modeling & Advancement in Research Trends (SMART) (pp. 410–414). IEEE.

Łapczyńska, D. (2023). The possibilities of improving the human–machine co-operation in semi-automatic production process. *Technologia i Automatyzacja Montażu (Assembly Techniques and Technologies), 119*(1), 30–36.

Longo, F., Padovano, A., & Umbrello, S. (2020). Value-oriented and ethical technology engineering in industry 5.0: A human-centric perspective for the design of the factory of the future. *Applied Sciences, 10*(12), 4182.

Maddikunta, P. K. R., Pham, Q. V., Prabadevi, B., Deepa, N., Dev, K., Gadekallu, T. R., ... & Liyanage, M. (2022). Industry 5.0: A survey on enabling technologies and potential applications. *Journal of Industrial Information Integration, 26*, 100257.

Mekkunnel, F. (2019). Industry 5.0: man–machine revolution (Doctoral dissertation, Wien).

Mourtzis, D., Angelopoulos, J., & Panopoulos, N. (2023). The future of the human–machine interface (HMI) in Society 5.0. *Future Internet, 15*(5), 162.

Nixdorf, S., Ansari, F., & Schlund, S. (2022). *Reciprocal learning in human–machine collaboration: A multi-agent system framework in Industry 5.0.* ID:363415836(pp. 207–225). Reasearch Gate Publication.

Ordieres-Meré, J., Gutierrez, M., & Villalba-Díez, J. (2023). Toward the industry 5.0 paradigm: Increasing value creation through the robust integration of humans and machines. *Computers in Industry, 150*, 103947.

Peruzzini, M., Prati, E., & Pelicciari, M. (2023). A framework to design smart manufacturing systems for Industry 5.0 based on the human-automation symbiosis. *International Journal of Computer Integrated Manufacturing.*

Pizoń, J., & Gola, A. (2023). Human–Machine Relationship – Perspective and Future Roadmap for Industry 5.0 Solutions. *Machines, 11*(2), 203.

Sindhwani, R., Afridi, S., Kumar, A., Banaitis, A., Luthra, S., & Singh, P. L. (2022). Can industry 5.0 revolutionize the wave of resilience and social value creation? A multi-criteria framework to analyze enablers. *Technology in Society, 68*, 101887.

Yang, J., Liu, T., Liu, Y., & Morgan, P. (2022, August). Review of human-machine interaction towards industry 5.0: human-centric smart manufacturing. In *International Design Engineering Technical Conferences and Computers and Information in Engineering Conference* (Vol. 86212, p. V002T02A060). American Society of Mechanical Engineers.

Zhang, C., Wang, Z., Zhou, G., Chang, F., Ma, D., Jing, Y., ... & Zhao, D. (2023). Towards new-generation human-centric smart manufacturing in Industry 5.0: A systematic review. *Advanced Engineering Informatics, 57*, 102121.

Chapter 9

Deploying new IT tools in Industry 5.0 in the creative direction

*Shashi Kant Gupta, Joanna Rosak-Szyrocka,
Chandra Kumar Dixit, Shovona Choudhury, and
Julee Banerji*

9.1 INTRODUCTION

It is crucial to adopt new information technology tools to transform and create value in the dynamic Industry 5.0 environment. The integration of cutting-edge information including blockchain technology, the Internet of Things (IoT), and artificial intelligence (AI) has caused a tremendous shift in the industry. The authors Carayannis et al. (2021) and Cillo et al. (2022) state that this change highlights the importance of being competent and effective, and creating significant advantages for consumers.

Technological improvements generated by Industry 5.0's usage of innovative IT technologies will considerably benefit manufacturing processes, supply chain management, and customer interaction, among many others (Cillo et al., 2022). The current focus is on developing a value cable that can adjust to new conditions via technology, rather than implementing automation and reducing costs. Zizic et al. (2022) and Sindhwani et al. (2022) state that companies are being encouraged to prioritize total value improvement over just turning a profit.

9.2 METHODOLOGY

All of the provided content on creativity is being examined thoroughly. An innovative method of managing the creative process is the focus of this undertaking. The suggested approach is beneficial for both administrators and consumers as it prioritizes simplicity, practicality, and the beneficial impacts of the technology. The best method for quickly creating new information in the innovation domain was determined by the research of Bartoloni et al. (2022). The purpose of the research by Aquilani et al. (2020) was to determine what factors prevent the widespread use of trendy innovation strategies. The study's overarching goal is to provide a revolution-ready framework compatible with IoT and Industry 5.0 requirements. Based on previous research (ElFar et al., 2021; Doyle-Kent and Kopacek,

DOI: 10.1201/9781032677040-9

2020), it was determined that a comprehensive methodology would be the best approach for this examination.

Research by Martynov et al. (2019) highlights the changing and dynamic nature of context management in the context of Industry 5.0 and the IoT. By omitting Industry 1.0 (the industries of the 19th century), Industry 2.0 (the Second Industrial Revolution), and Industry 3.0 (the Digital Revolution), we may follow the development of Industry 5.0 through its successive stages. The IoT has recently seen tremendous advancements, leading to the rise of Industry 4.0, often known as the Fourth Industrial Revolution. Various technological developments including "IoT, Big Data, electric vehicle technology, three-dimensional printing, cloud computing, and artificial intelligence" (Marinič and Pecina, 2023; Carayannis et al., 2021), have been integrated into the current age, which began in 2000 and continues until the present day. From 2016 onwards, the concept of Industry 5.0 has been gaining traction alongside Industry 4.0 (Wang et al., 2023). The emergence of a society that is highly technologically advanced is what defines Industry 5.0. Scholarly publications spanning nearly two decades, from 2000 to 2019, were purposefully included due to the rise of Industry 4.0 and 5.0 theories and the improvements in the IoT (Sołtysik-Piorunkiewicz and Zdonek, 2021).

9.3 THE REQUIRED FRAMEWORK FOR NEW INNOVATION STRUCTURE

The implementation ability of a new idea is the most crucial feature. The concept of implementing ability pertains to the power of new technology to provide value for its clients or consumers. If a new idea fails to generate value for consumers or users, it becomes unsustainable for implementation (Carayannis et al., 2021). If a new idea proves insufficient in fulfilling human requirements, failing to benefit customers or users, it becomes imperative to continue the novel procedure. The significance of making consumers or end-users the center of attention throughout the innovation process must be fully appreciated (Salimova et al., 2019). Innovation, according to this logic, must center on people's wants and needs if it is to provide value for consumers. One of the guiding principles of Industry 5.0 is the idea that human needs and preferences should take precedence (Mourtzis et al., 2022). Moving beyond digital production and toward a digital society is an integral part of the progression toward Industry 5.0. A new innovation framework, one that prioritizes the creation of value for businesses and their customers alike, is required to facilitate this shift. The Fifth Industrial Revolution, or Industry 5.0, is defined as a highly technologically sophisticated society that merges the digital and physical realms. Robotics, augmented reality, the IoT, and an innovation system are all part of it. Furthermore, it highlights the importance of brain–machine interfaces and how they might involve humans in technological progress. Carayannis et al. (2021), SoŚtysik-Piorunkiewicz and Zdonek (2021), and Saniuk et al. (2022) have all addressed these ideas. Consumers and business owners are the intended beneficiaries of these products. Prioritizing user needs over technological advancements should guide the development of IoT devices and infrastructure. According to Rožanec et al. (2023) and Javaid and Haleem (2020), developing "people-aware" apps for IoT devices and products requires an inventive mindset in addition to appropriate methodology, tools, and methodologies. Invention should prioritize the needs and preferences of individuals and users while producing value for both customers and commercial owners (Khan et al., 2022).

The development of novel models represents significant developments, with each structure attempting to respond to the existing requirements and challenges of the respective eras. The presence of novelty changes and limited comprehension of the novel impressions have presented challenges for novel models (Saniuk et al., 2022). This is because many current innovation models and structures have been designed to function within stable and well-established contexts, where organizations focus on minor improvements in their operations compared to their competitors (ElFar et al., 2021; Martynov et al., 2019; Carayannis et al., 2021). These changes could push administrations to operate elsewhere their established processes and structures when constant usage of the same actions no longer produces favorable outcomes. Each subsequent innovation model has been developed by addressing the limitations and defects of previous invention models. The creation of innovation control (IC) was inspired by the principles outlined in the fifth-generation model, with a specific emphasis on creating a favorable atmosphere for invention (Rožanec et al., 2023).

To achieve the goal of creating creations that select the needs of both customers and business owners, and to seamlessly incorporate this process into the everyday actions of a business rather than considering it as a separate and inaccessible activity, this segment introduces a novel framework for innovation. The contextual has been described as "innovation" in the works of Cillo et al. (2022), Doyle-Kent and Kopacek (2020), Mourtzis et al. (2022), and Javaid and Haleem (2020).

9.4 INNOVATION CONTROL

This study suggested a new framework called creativity management to address the issues acknowledged in previous studies. The purpose of IC is not to create an entirely new approach but rather to address the limitations present in current innovation structures. The IC included specific components from pre-existing innovation structures while introducing novel features as considered essential.

Definition of IC: The IC structure refers to a conceptual structure that establishes a connection between the innovation environment and the overall strategy of a firm. It is achieved by implementing IC as a strategic approach, utilizing design thinking principles to ensure that innovation is centered on the needs and preferences of users or humans. The objective is to create innovations that are desirable to customers, profitable for the business, and technically achievable. To enhance business ventures and business owners while generating entrepreneurial and customer value, it is imperative to pursue sustainable superiority and economic growth. This approach should address the demands of Industry 5.0 and the era of the IoT.

Basic Components of IC: Based on an extensive examination of existing academic articles on innovation structures with consideration for the expected requirements of "the IoT and Industry 5.0 era" (Khan et al., 2022), this study proposes an innovation structure that consists of three fundamental components as follows:

- Innovation environment
- Creative thinking
- Strategy for IC

Innovation Environment: The primary basic component of innovation capability is the innovation environment (Ivanov, 2023). The innovation environment inside the IC structure encompasses the attributes of fifth-generation innovation models.

The concept of the innovation environment asserts that a conducive environment is necessary for encouraging innovation, hence yielding benefits for everyone involved. The innovation environment includes three overarching components, namely the integration of innovation across all functions, encompassing technical and non-technical dimensions. The second aspect relates to the cultivation of innovation among all members of the organization, while the third aspect emphasizes the importance of fostering innovation across various temporal and geographic dimensions (Carayannis et al., 2021; Martynov et al., 2019; Ivanov, 2023). The innovation environment clarifies that invention extends beyond technical domains (George and George, 2020). The innovation environment distributes accountability for innovation throughout the company by asserting invention as a collective obligation, irrespective of one's role or hierarchical position.

The innovation environment exhibits a propensity for fostering creativity in many temporal and spatial contexts. "All spaces" refers to the encompassing scope of innovation inside and across the company, encompassing inter- and intra-firm connectivity as well as interaction. "By all times" refers to the understanding that innovation is neither a singular nor an isolated endeavor, but rather an ongoing and continuous process (Zambon et al., 2019). Therefore, it is essential for organizations to be involved in the practice of innovation to develop flexibility (Khan et al., 2022; Ivanov, 2023).

Creative Thinking: Implementing creative thinking as part of an innovation structure is central to the IC approach. While creative thinking is recognized in innovation, especially in IT and other fields, the previous structure must be addressed or incorporated. Effective use of creative thinking in innovation is crucial for surviving the "IoT and Industry 5.0 era." The critical difference between IC and the earlier structure is creative thinking, making it more advanced.

This concept originated from the "Organization for Economic Co-Operation and Development (OECD)" and states that for an innovation to be considered innovative, it must be able to be implemented into practice and provide some benefit to those who will be using it. Following the delineation of the terrible issue described by sources (Salimova et al., 2019; Carayannis et al., 2022), innovation is classified as a wicked problem by its inherent attributes. Hence, compared to an established challenge, the nature of innovation as an unjust issue requires a distinct and unconventional approach in terms of cognitive processes and methodologies for its resolution. Creative thinking has been identified as a potential solution to the complex and challenging issue of innovation (Wang et al., 2023; George and George, 2020).

Applying creative thinking to address the complicated issue of innovation requires establishing an appropriate mental model and framing of the circumstances (Bartoloni et al., 2022). The imperfect understanding of invention discussed in previous sections (Aquilani et al., 2020; Wang et al., 2023; Javaid and Haleem, 2020; Chen et al., 2023) can be addressed through innovative thinking, cognition frameworks, and the contextualizing of unique situations. The basic concept of creative thinking is summarized in Figure 9.1.

Strategy for IC: Strategy is the third and final element of IC. The innovation environment is highlighted more in IC (Salimova et al., 2019), but there needs to be more balance between the environment and business strategy. While Mourtzis et al. (2022) attempted to link the innovation environment to the established strategic objective and structure, the standard logic of the firm will oppose this linkage, making this strategy inappropriate for an IoT and Industry 5.0-dominated world. In addition, earlier innovation structures lacked the adhesive necessity for connecting the innovation environment with business objectives (Yosintha, 2020).

Innovation correlates with the concepts of "corporate venturing" and "corporate business ownership," which face resistance from established "regular logic." In essence, the customary activities of a corporation present challenges for business ownership and corporate owners (Wang et al., 2023; Mourtzis et al., 2022; Khan et al., 2022). To introduce human-centered innovation into an organization's normal operations while ensuring commercial

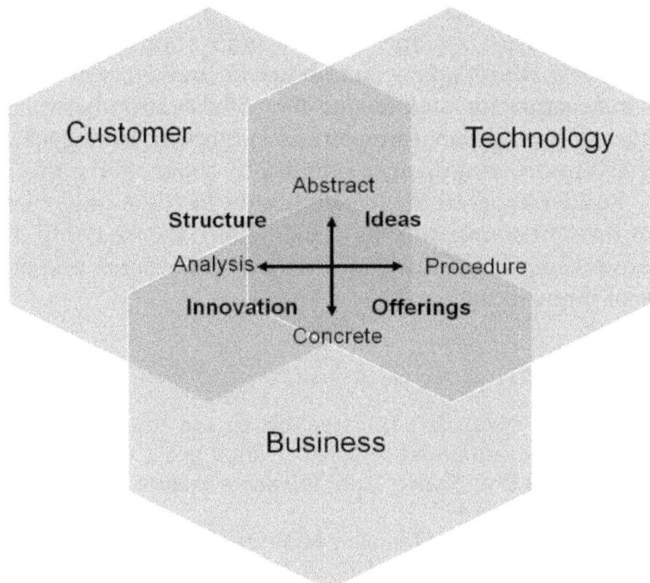

Figure 9.1 Creative thinking.

viability, it is advisable to view innovation as a strategic approach rather than a mere supportive tool (Doyle-Kent and Kopacek, 2020).

Each organization has the authority to determine its approach to incorporating invention as a strategy and to find a harmonious equilibrium between innovation and established methods. It is imperative to underscore that the fundamental nature of this idea is in a cognitive transformation, irrespective of the chosen approach. Approval of this perspective was provided by Yosintha (2020), who asserted that a universal system capable of accommodating all organizations and functioning in all situations does not exist. Instead, embracing innovation as a strategic approach enables firms to develop a strategy that aligns with their unique competitive requirements (Carayannis et al., 2021; Cillo et al., 2022; Aquilani et al., 2020; George and George, 2020). Implementing customization, when considered a strategic approach to innovation and recognized as a dynamic concept, has the potential to yield more favorable outcomes.

9.5 CONCLUSION

This investigation has made multiple contributions to the literature. To begin, the organization developed an IC in an effort to remove the confusion and inadequacy that have clouded discussions of the process of implementing innovation. Secondly, the IC structure encourages participation from the company members, which not only speeds up the innovation process but also instills a feeling of ownership and accountability in employees at all levels. Finally, it offers a structure for maximizing the collaboration between creative thinking, the innovation environment, and company strategy. Rather than controlling inventions as a distinct activity, IC enhances the invention part of ordinary business operations, which allows for "the implementing ability of innovation to be simple and smooth." Finally, the idea of IC assists in reducing the adverse outcomes associated with discontinuities and incomplete knowledge of innovation.

REFERENCES

Aquilani, Barbara, Michela Piccarozzi, Tindara Abbate, and Anna Codini. 2020. "The role of open innovation and value co-creation in the challenging transition from Industry 4.0 to Society 5.0: Toward a theoretical framework." *Sustainability 12*, no. 21: 8943.

Bartoloni, Sara, Ernesto Calo, Luca Marinelli, Federica Pascucci, Luca Dezi, Elias Carayannis, Gian Marco Revel, and Gian Luca Gregori. 2022. "Towards designing Society 5.0 solutions: The new Quintuple Helix-Design Thinking approach to technology." *Technovation 113*: 102413.

Carayannis, Elias G., David F. J. Campbell, and Evangelos Grigoroudis. 2022. "Helix trilogy: The triple, quadruple, and quintuple innovation helices from a theory, policy, and practice set of perspectives." *Journal of the Knowledge Economy 13*, no. 3: 2272–2301.

Carayannis, Elias G., Luca Dezi, Gianluca Gregori, and Ernesto Calo. 2021. "Smart environments and techno-centric and human-centric innovations for Industry and Society 5.0: A quintuple helix innovation system view towards smart, sustainable, and inclusive solutions." *Journal of the Knowledge Economy 13*: 926–955.

Carayannis, Elias G., John Draper, and Balwant Bhaneja. 2021. "Towards fusion energy in the Industry 5.0 and Society 5.0 context: Call for a global commission for urgent action on fusion energy." *Journal of the Knowledge Economy 12*, no. 4: 1891–1904.

Chen, Xiao, Martin Alexander Eder, and A. S. M. Shihavuddin. 2023. *A Concept for Human–Cyber–Physical Systems of Future Wind Turbines Towards Industry 5.0.* Authorea Preprints.

Cillo, Valentina, Gian Luca Gregori, Lucia Michela Daniele, Francesco Caputo, and Nathalie Bitbol-Saba. 2022 "Rethinking companies' culture through knowledge management lens during Industry 5.0 transition." *Journal of Knowledge Management 26*, no. 10: 2485–2498.

Doyle-Kent, Mary, and Peter Kopacek. 2020. "Industry 5.0: Is the manufacturing industry on the cusp of a new revolution?" In *Proceedings of the International Symposium for Production Research 2019*, pp. 432–441. Springer International Publishing.

ElFar, Omar Ashraf, Chih-Kai Chang, Hui Yi Leong, Angela Paul Peter, Kit Wayne Chew, and Pau Loke Show. 2021. "Prospects of Industry 5.0 in algae: Customization of production and new advance technology for clean bioenergy generation." *Energy Conversion and Management X 10*: 100048.

George, A. Shaji, and A. H. George. 2020. "Industrial Revolution 5.0: the transformation of the modern manufacturing process to enable man and machine to work hand in hand." *Journal of Seybold Report* ISSN No. 1533: 9211.

Ivanov, Dmitry. 2023. "The Industry 5.0 framework: Viability-based integration of the resilience, sustainability, and human-centricity perspectives." *International Journal of Production Research 61*, no. 5: 1683–1695.

Javaid, Mohd, and Abid Haleem. 2020. "Critical components of Industry 5.0 towards a successful adoption in the field of manufacturing." *Journal of Industrial Integration and Management 5*, no. 03: 327–348.

Khan, Wali Ullah, Asim Ihsan, Tu N. Nguyen, Zain Ali, and Muhammad Awais Javed. 2022. "NOMA-enabled backscatter communications for green transportation in automotive-industry 5.0." *IEEE Transactions on Industrial Informatics 18*, no. 11: 7862–7874.

Martynov, Vitaly V., Diana N. Shavaleeva, and Alena A. Zaytseva. 2019. "Information technology as the basis for transformation into a digital society and Industry 5.0." In *2019 International Conference on Quality Management, Transport and Information Security, Information Technologies (IT&QM&IS)*, pp. 539–543. IEEE.

Marinič, Peter, and Pavel Pecina. 2023. "Industry 5.0 in vocational education." In *INTED2023 Proceedings*, pp. 3805–3811. IATED.

Mourtzis, Dimitris, John Angelopoulos, and Nikos Panopoulos. 2022. "A literature review of the challenges and opportunities of the transition from Industry 4.0 to Society 5.0." *Energies 15*, no. 17: 6276.

Rožanec, Jože M., Inna Novalija, Patrik Zajec, Klemen Kenda, Hooman Tavakoli Ghinani, Sungho Suh, Entso Veliou, et al. 2023. "Human-centric artificial intelligence architecture for industry 5.0 applications." *International Journal of Production Research 61*, no. 20 (2023): 6847–6872.

Salimova, Tatyana, Nadezhda Guskova, Irina Krakovskaya, and Efim Sirota. 2019. "From Industry 4.0 to Society 5.0: Challenges for the sustainable competitiveness of Russian industry." In *IOP Conference Series: Materials Science and Engineering*, 497: 012090. IOP Publishing.

Saniuk, Sebastian, Sandra Grabowska, and Martin Straka. 2022. "Identification of social and economic expectations: Contextual reasons for the transformation of Industry 4.0 into the Industry 5.0 concept." *Sustainability 14*, no. 3: 1391.

Sindhwani, Rahul, Shayan Afridi, Anil Kumar, Audrius Banaitis, Sunil Luthra, and Punj Lata Singh. 2022. "Can Industry 5.0 revolutionize the wave of resilience and social value creation? A multi-criteria framework to analyze enablers." *Technology in Society 68*: 101887.

Sołtysik-Piorunkiewicz, Anna, and Iwona Zdonek. 2021. "How Society 5.0 and Industry 4.0 ideas shape the open data performance expectancy." *Sustainability 13*, no. 2: 917.

Wang, Xingxia, Jing Yang, Yutong Wang, Qinghai Miao, Fei-Yue Wang, Aijun Zhao, Jian-Ling Deng, Lingxi Li, Xiaoxiang Na, and Ljubo Vlacic. 2023. "Steps toward industry 5.0: Building "6S" parallel industries with cyber-physical-social intelligence." *IEEE/CAA Journal of Automatica Sinica 10,* no. 8: 1692–1703.

Yosintha, Rolisda. 2020. "Indonesian students' attitudes towards EFL learning in response to Industry 5.0." *Metathesis: Journal of English Language, Literature, and Teaching 4,* no. 2: 163–177.

Zambon, Ilaria, Massimo Cecchini, Gianluca Egidi, Maria Grazia Saporito, and Andrea Colantoni. 2019. "Revolution 4.0: Industry vs. agriculture in a future development for SMEs." *Processes 7,* no. 1: 36.

Zizic, Marina Crnjac, Marko Mladineo, Nikola Gjeldum, and Luka Celent. 2022. "From Industry 4.0 towards Industry 5.0: A review and analysis of paradigm shift for the people, organization, and technology." *Energies 15,* no. 14: 5221.

Chapter 10

Quality education in the era of Industry 5.0

Exploration of the role of educational leaders and managers for value creation

Zafarullah Sahito, Noor Ahmed, Anna-Marie Pelser, and Insaf Ali

10.1 INTRODUCTION

Education is a container of the thoughts to transmit knowledge (Hooks 2003; Freire 2018), skills, values, and ethics (Miller and Miller 2000; Warnock 1986) to the members of the society for its betterment (Halpern 2010; Strickland and Kaylor 2016), as the individuals can live peacefully (Miller and Miller 2000). The individuals who make society peaceful through socialization or enculturation of education are known as social scientists (Dewey 1916; Freire 2018); they guide society to learn a culture in which behavior can be molded among adults to direct them to explore different avenues (Lieberman 2015) to play their ultimate role in society for their personal and others benefits (Halpern 2010). The quantity of knowledge passed from one generation to the next becomes more than a person's knowledge in any multifaceted societies, because of the advancement and distinction in systems. The advancement in systems also brought the modification in the methods and the materials, techniques, and strategies (Dewey 1938) to deal with and fulfill the needs and requirements (Lieberman 2015; Smith 2008) of the modern world in the 21st century. Education states the discipline(s) concerned with the methods and styles of teaching and learning in different institutions and environments (Hooks 2013) but not specify in advance, i.e., schools, colleges, and universities known as formal and non-formal institutions (Ellis 2021) because there are some limitations of the world (Halpin and Halpin 2002). Therefore, the limitations play a vital role in creating some systems of education with some hopes and aims to fabricate the education (Warnock 1986), i.e., there are three systems of education, namely, formal, non-formal, and informal, where any individual can get education and learn the knowledge and skills (Mowen 2013) to use in their day-to-day life to be a successful human being in the society with the implementation of knowledge, concepts, ideas, beliefs, and visions (Mowen 2013). Where, socialization can be the on the priority to interact with the things, systems, individuals, etc. to learn scientifically to be a useful human resources.

DOI: 10.1201/9781032677040-10

Learning through different styles, ways, and systems is directly connected with the three main types of education, i.e., formal, informal, and non-formal (Curtis et al. 2013; Eshach 2007). Whereas each system has its characteristics, strengths, weaknesses, advantages, and disadvantages, it directly and indirectly, supports the individuals to get education and learn new things (Crowley 2014; Ullah et al. 2012) scientifically in every condition of life, status, position, etc. but a majority of theorists distinguished between formal and informal education (Strauss 1984). A formal education system takes place in a complex institutionalized framework with chronological and hierarchical order starting from primary to university level organizations, guided by governmental instruction(s) with certain compulsory age limits (Tudor 2013), while non-formal and informal education systems vary from formal education due to lack of governmental institutionalized framework, fixed place of learning, freedom, time limit, etc. Non-formal education constitutes a middle ground between formal and informal with an organized and systematic sort of balance to carry out everything with a clear purpose, i.e., tutoring, assessing, etc. (Curtis et al. 2013; Tudor 2013), finally and occasionally visits are to be arranged in non-formal education (Eshach 2007; Tudor 2013), where the students' motivation is made responsible and works intrinsic to a non-formal system of education (Eshach 2007; Tudor 2013). The difference between all three types of education systems depends on the exemplary cases that support the open education system (Marín et al. 2020) to encourage individuals to learn as per their freedom, interests, wishes, and will.

10.2 OBJECTIVES AND THE RESEARCH QUESTIONS OF THE CHAPTER

The objectives of the chapter are as follows: (a) To understand the concepts of Industrial Revolutions from Industry 1.0 to 5.0; (b) to understand the concept of Quality Education, and Quality Assurance; and (c) to discuss the Role of Educational Leaders and Managers in maintaining Quality Education in the era of Industry 5.0. In the light of objectives, the research questions of the chapter are as follows: (a) What is Industrial Revolution? (b) How did Industry Revolution 1.0 to 5.0 take place? (c) What is Quality Education and Assurance? (d) What is the role of Educational Leaders and Managers in the maintenance of Quality Education?

10.3 INDUSTRIAL REVOLUTIONS

The First Industrial Revolution started from handmade production and methods to machines, especially the use of steam power and water power. The implementation period of technologies was referred to 1760 to 1820, or 1840 in Europe and the USA. In the 18th century, mechanical manufacturing was piloted in the industrial era. Machines and engines transformed the way things were manufactured; likewise, James Watt's discovery of the steam engine was a main transition from an agrarian to an industrial civilization, which was accelerated by the advent of the steam engine in England (Colangelo and Bauernhansl 2016). It directly affected the textile industry, which was the first industry to adopt the changes, and then the iron, agriculture, and mining industries, which had a great effect on the society, especially the society and industry of the United Kingdom (Cannadine 2003).

In the 19th century, Industrial Revolution 2.0 (IR 2.0) was started in Europe and the USA as a revolution (Cannadine 2003), which was distinguished by mass manufacturing and the substitution of chemical and electrical energy for steam (Colangelo and Bauernhansl 2016). To satisfy the continued growth, many industrial and mechanization technologies have been created to increase production. This Second Industrial Revolution was also known as the Technological Revolution between 1871 and 1914 (Cannadine 2003). As a result, many new things have been achieved, i.e., installations and construction of widespread railway tracks, roads, and telegraph networks to fasten the transfer of people, goods, and ideas. Electricity was a prominent invention and achievement in human history, which increased the electrification of factories to develop the modern and quick production of many items to increase productivity for economic growth and development through technology and reduce unemployment (Hermann, Pentek, and Otto 2016).

The Third Industrial Revolution, also known as the Digital Revolution, started in the late 20th century (Lasi et al. 2014), which was the result of two world wars, i.e., slowdown of industrialization and technological advancement compared to previous years (Yin et al. 2018). The production of computers, i.e., binary floating-point numbers and Boolean logic were found more advanced digital developments, whereas the most prominent and significant development in the field of communication technologies was the supercomputer. Computers and other communication technologies were used extensively for production process, i.e., machinery, etc. to increase human power (Longo et al. 2021). IR 3.0 was started with the development of the integrated circuit (microchip), which was a great technological breakthrough, started to utilize technology to accomplish automation in manufacturing with a significant characteristic of this revolution to arose in many developed nations throughout the world in the later years of the 20th century (Acemoglu and Linn 2003; Von Tunzelmann 2003).

The Fourth Industrial Revolution was first introduced by a team of scientists, who were developing a high-tech strategy for the German government at the Hannover Fair in 2011. The concept prevailed worldwide due to its strategy based on the newly organized operations related to industries through new technologies. The executive chairman of the World Economic Forum (WEF) introduced the phrase to a wider group of audience in 2015, which was published in an article by Foreign Affairs. The mastering on the Fourth Industrial Revolution was done in the year of 2016 and presented in the shape of a theme during the annual meeting of the World Economic Forum in Davos-Klosters, Switzerland (Marr 2015). Afterward, on 10th October 2016, the WEF announced the opening of its new center to work for the Fourth Industrial Revolution in San Francisco, which was the title and the subject of the book by Schwab published in 2016. In this book, Schwab includes and discusses the fourth-era technologies to combine hardware, software, and biology known as cyber–physical systems (CPS) where the writer focuses his emphasis on the advancement of communication and connectivity to minimize the distances for better results, because the writer considers this era to be marked as a breakthrough in the fields of emerging technologies in fields of robotics, artificial intelligence, nanotechnology, quantum computing, biotechnology, the Internet of Things (IoT), the industrial IoT, decentralized consensus, fifth-generation wireless technologies, three-dimensional (3D) printing, and fully autonomous vehicles.

Furthermore, in the proposal of WEF, the Fourth Industrial Revolution was included as a Strategic Intelligence to bring solutions for rebuilding the economy with sustainability before and after any pandemic, i.e., COVID-19. The Fourth Industrial Revolution is an advanced step towards the computerization and data exchange of manufacturing technologies and processes, i.e., CPS, IoT, industrial IoT, cloud computing, cognitive computing, and artificial intelligence (Lasi et al. 2014; Longo et al. 2021) that characterizes the beginning of the Imagination Age. The nine pillars of Industry 4.0 have been defined by Boston Consulting Group, which are the latest technologies known by the world: (a) Multi-Agent Systems, an intelligent smart machines, collaborating robots, sensors, controllers, etc., can be used for excellent types of communication through autonomous robots and artificial intelligence; (b) System integration, the optimal reconfiguration of the connecting CPS, works as horizontal integration; (c) Big data refers to the huge amount of information required for the optimal operations to make real-time decisions; (d) Simulation tools are used for optimization of production processes and maximal utilization of resources; (e) Cyber security comprises technologies that are developed to protect systems, networks, and data from cyber-attacks; (f) Cloud computing offers unlimited computing power to transfer, store, and analyze a huge amount of data; (g) Additive manufacturing is an innovative technology to build 3D objects like 3D printing technology that provides the possibility of manufacturing more complex and unique products; (h) Augmented reality provides the possibility of visualization of manufacturing processes through transforming from the real environment to a virtual environment; (i) The IoT is a technology that allows objects, i.e., machines, vehicles, products, etc. to communicate and interact with each other (Vaidya et al. 2018).

Industry 5.0 (IR 5.0) refers to robots and smart machines helping human beings and working together for the betterment of society through faster leveraging advanced technologies, i.e., IoT (Lampropoulos et al. 2019), artificial intelligence (Drexler and Lapré 2019), big data (Lampropoulos et al. 2018), etc. The digital evolution of the last decade has established the groundwork for the era of human and machine collaboration, known as Industry 5.0 (Musarat et al. 2023). This era focuses on enhancing collaboration between human beings and intelligent systems to address mass personalization and improve work efficiency (Concepcion et al. 2022). George and George (2020) termed IR 5.0 as a set of revolutionizing the manufacturing sector, acting as a catalyst for an industrial revolution. Ideally, IR 5.0 represents the evolution of modern manufacturing processes, enabling human beings and machines to collaborate closely, to leverage the distinctive cognitive skills of human workers and the precise technical expertise of robots to foster a culture of innovation in the workforce. Furthermore, Humayun (2021) emphasized that IR 5.0 is primarily pushed by the consumer demand for customized and personalized products, which calls for increased human participation in the manufacturing process. As IR 5.0 advances, we can anticipate numerous innovations across multiple industries. Nevertheless, merely automating tasks or digitizing processes will not suffice; the most exceptional and prosperous businesses will be the capable system of harmonizing the twin forces (Musarat et al. 2023), i.e., technology and human creativity. One of the prime advantages of IR 5.0 is its commitment to environmentally friendly solutions, in contrast to the traditional industrial transformations that often do not prioritize environmental protection (Maddikunta et al. 2022; Show et al. 2021). As per the insights of the co-founder and Chief Technology Officer of Universal Robots, IR 5.0 is set to reshape the factory environment into a space (George and George 2020) where creative individuals can collaborate and foster a more personalized and human-centric experience for both workers and consumers (Taj and Zaman 2022), because the Fifth Industrial Revolution may bring more significant and intricate collaborative interactions to involve the intellect, machinery, processes, and the overall system to achieve optimal performance optimization (Majid et al. 2019).

The continuous advancement of technology is reshaping our daily lives, making them intelligent and convenient (Diamandis and Kotler 2020), i.e., Industry 5.0, Society 5.0, Healthcare 5.0, and Agriculture 5.0, etc. to illustrate the rapid transformation of our lifestyle to create a smart future (Lee and Trimi 2018). IR 5.0 encompasses the forthcoming industry development trends that aim to achieve prosperity that goes beyond traditional employment, which can be achieved by integrating greater intelligence into our daily lives through cutting-edge technologies like artificial intelligence (Lee 2020), etc. There is no doubt that the history of the Industrial Revolution's progress has had a profound influence on the advancement of the global economic sector. As Purnamasari et al. (2019) stated, the emergence of Industrial Revolution 4.0 (IR 4.0) marked the complete digitalization in the industrial sector and has represented a significant development since 2018, whereas IR 5.0 offers the potential to address societal challenges, including the depletion of natural resources and economic inequality on a global scale. Taj and Zaman (2022) defined the term IR 4.0 as pertaining to the technological advancements that introduce new ways of living, leading to a profound transformation of society.

10.4 QUALITY EDUCATION

Quality of education is a quantifying process of the effectiveness (Dello-Iacovo 2009; Kumar and Manivannan 2023) of a daily routine in any educational institution (Nussbaum 2006) to meet their vision and mission through the achievement of predetermined goals (Biggs et al. 2022). Most organizations consider the quality of education as the insurance of lifelong learning opportunities (Visvizi et al. 2020) for all the members of society without any difference or discrimination. It helps to end inequality through the termination of the cycle of poverty to provide quality life opportunities to the concerned members and stakeholders (Pramana et al. 2021), because quality education leads people to a healthy, happy, joyful, and sustainable life to live peacefully in their communities and societies (Visvizi et al. 2020). The major approaches to maintaining quality education are the proper understanding and the usage of (a) philosophy; (b) the product-based approach of economics; (c) the user-based approach of economics, marketing, and operations management; (d) the manufacturing-based approach; and (e) value-based approaches to operations (Jones 2014; Rosak-Szyrocka et al. 2023). The standards of quality education are focusing on the areas of improvement, i.e., upgradation and improvement of curricula; writing and publication of quality textbooks; creation of the pool of trained and dedicated teachers; designing of sound internal and external assessment practices; and the creation and maintenance of a conducive learning environment inside and outside the classrooms and institutions (Mitra 2016).

The quality assurance system becomes an area of strategic plan among top-ranking institutions (Didou 2022; Hou et al. 2012; Mourato et al. 2021) to ensure the learning outputs for the confirmation of quality education (Marciniak 2018) through the development of quality assurance criteria and system (Kishabale and Hassan 2018) to support all stakeholders (Mahdiuon et al. 2017) to achieve the predetermined goals (Nayyar and Kumar 2020) of the components of quality education, learning (Nayyar and Kumar 2020), and learning environments (Marciniak 2018). Quality education is a prioritized and emerging area of the institutions of the globe (Avota 2018; Batsila 2020; Li et al. 2022; Taftaf and Williams 2019; Scheunemann et al. 2018; Stambekova et al. 2022; Zawacki-Richter and Qayyum 2019), where the subject matter and learning outcomes are assessed regularly (Mukhopadhyay et al. 2020) to sustain quality education (Pandey and Indrakanti 2017) to enhance the educational experiences of the learners (Vlachopoulos and Makri 2019). Quality education further focuses on the pedagogies (Al-Hosni and Al-Ali 2022), technologies (Al-Sharhan et al. 2016; Holsombach-Ebner 2013; Nsiah and Oti-Boadi 2015), assessment (Bates et al. 2016), and mechanisms (Nsiah and Oti-Boadi 2015) because quality assurance is a systematic management and assessment of the procedures to monitor the performance for improvements (Belawati and Zuhairi 2007) to support student learning (Mukhopadhyay et al. 2020), which can be achieved through the enhancement of the roles and responsibilities of the students.

Quality helps individuals and organizations to boost their reputation, value of their brand(s) and meet the needs, requirements, and standards of the industry (Litvaj et al. 2023). The quality control system of any organization is a product-based process that makes the assurance of its quality products increase and enhance its name and fame in the market (Baig 2023). The importance of quality and its culture lies in lowering employee turnover, motivating them to participate actively to increase and improve performance and product level through their conducive working conditions, environment, employees' health, and well-being (Budiharso and Tarman 2020). Therefore, educational leaders and managers need to be aware of the emerging challenges, issues, and problems with scientific solutions in the age of rapid change (Tavares et al. 2022) like IR 5.0 (Carayannis and Morawska-Jancelewicz 2022; Mohammadian 2020). The positive fulfillment of the roles and responsibilities of heads brings improvement and enhancement of quality education (Chinooneka 2020; Salendab and Dapitan 2021), which is the prime need and requirement of the 21st century. The heads are to be well-versed with the mechanisms and their tools to maintain quality education (Msuya and Mwila 2023) in the era of IR 5.0 where other countries are far ahead of developing countries. If the educational leaders and managers use a positive and implementable vision and mission to upgrade their existing systems through preplanning based on modern needs and requirements (Bafadal and Arifin 2020), then there will be a great possibility to have good opportunities to maintain quality education (Dube-Xaba and Makae 2022) through establishing of the effective and sustainable education system to achieve the millennium development goals (MDGs) (Nazar et al. 2018).

10.5 ROLE OF EDUCATIONAL LEADERS AND MANAGERS TO IMPROVE THE QUALITY OF EDUCATION IN THE ERA OF INDUSTRY 5.0

Educational leaders and managers are qualified experts who work for the betterment of their organizations (Day et al. 2016) to achieve the predetermined goals. They try their level best to tie talented people with their organizations to utilize the knowledge, and skills of the people to strengthen the system of the organization through a conducive environment to achieve the organizational goals for the collective benefit of the society. Educational organizations are known as social institutions, which create good leaders, managers, and human resources to support the systems of society. Qualified human resources are considered as strength of any organization that always try their level best to work innovatively to maintain discipline and progress in different forms like the quality education in educational institutions (Sahito and Chachar 2021). Quality education creates awareness and provides knowledge and skills among people to bring modification in the shape of change in society through their organizations and new concepts like Industry 5.0, purely based on equality, equity, and social justice. Furthermore, the citizens of the society are provided all basic rights to live a quality life within the available resources (Alfadala et al. 2021; White and McCallum 2020), especially education and technology. The prime aim of the leaders and managers is to ensure educational success through maintaining and managing the processes, planning things, developing materials, and conducting training for the betterment and improvements through acquiring the collaboration of all concerned stakeholders (Day et al. 2016; Leaver, Lemos, and Scur 2019). Further, educational leaders focus on the prime principles to support and improve the education and learning systems (Ezzani et al. 2021; Shakeshaft et al. 2014) through playing their active role.

The role of educational leaders and managers is described below:

1. To create an organizational vision, execute it through different programs, and connect it with future endeavors (Lane 2012) to support the sustainable development goals (SDGs).
2. To provide a sense of direction in the light of vision to prioritize the focus and the attention (Childress et al. 2020) required by Industry 5.0.
3. To minimize the gaps between all stakeholders based on different socio-economic conditions and achievement levels based on rules of equality, equity, and justice.
4. To struggle for a safe, secure, affordable, and approachable learning environment (Parkin 2022) for better performance.
5. To manage and maintain a conducive school and classroom environment for a comfortable, easy, and structured modernized system for all as needed by Industry 5.0.
6. To design the student-centered policies as per the need of the 21st-century education system, i.e., a student-centered system to compete globally with other independent nations.
7. To delegate powers, duties, and responsibilities to all concerned and make them accountable for easy accessibility to facilitate all across

the systems with the help of Industry 5.0 to maintain prosperity through quality.

8. To empower all the concerns as per their needs and requirement of their jobs to make them active and responsible team members through proper planning and decision-making to promote leadership skills among them (Gates et al. 2019) in the collaboration of the rules and regulations of the Industry 5.0.

9. To support and appreciate all the team members to review, revise, and upgrade their list of duties and responsibilities with needed improvements and incorporation (Alfadala et al. 2021) to achieve the aims and objectives of Industry 5.0.

10. To appreciate all the colleagues to derive, acclimate, and employ the management tools, techniques, rules, regulations, and best practices of the world to improve the system of quality education (Gates et al. 2019) as needed in Industry 5.0.

11. To build the community and its partnership for sustainable common enterprises for better development of understanding, trust, and business through transparency in the era of Industry 5.0.

12. To be passionate about their work and the expected results, which may be achieved through some risk-taking (Jensen et al. 2017), which is the need of the hour and smart people aligned with Industry 5.0.

13. To appreciate all the concerned stakeholders to learn well and be the lifelong learner moving with the futuristic approach of the smart people using smart technology in Industry 5.0.

14. To make all systems efficient and effective through active engagement of all stakeholders to make them satisfied and motivated through designing comprehensive programs to make smart and efficient systems (Gates et al. 2019), which is the ominous need of the era of Industry 5.0.

15. To make all stakeholders familiar, simple, and flexible to organize the systems of fulfillment of duties and responsibilities to maintain and enhance the quality of the education system (Gray 2018) in the era of Industry 5.0.

16. To install efficient systems to check the regularity and punctuality of all stakeholders to maintain the quality of education (Breakspear 2017; Giesbers et al. 2019), because quality education can be connected easily through the strong bond among the stakeholders and systems.

17. To increase and enhance the capability of all the stakeholders to choose and decide the multiple modes of data for better decisions (Kesler et al. 2022; Phelps 2008) based on the asynchronous and synchronous systems, which are highly required for Industry 5.0.

18. To focus on the learning of modern pedagogical techniques of all stakeholders to use project-based learning (Kesler et al. 2022), experiential learning, etc. with the freedom to understand the needed information and phenomenon of Industry 5.0.
19. To provide professional growth and development facilities to all concerned to get maximum benefits (Onwujekwe et al. 2022) for maintaining quality as is the requirement of all systems like Industry 5.0.
20. To provide quick and adequate support to all concerned to study, and research the existing issues and problems in their related areas and collect the empirical pieces of evidence of the proper and scientific solutions (Onwujekwe et al. 2022) to achieve aim of the Industry 5.0.
21. To focus on the funding increase to bring financial sustainability (Yun et al. 2019) to the organization (Ahmada and Fakhr 2022), which provides safety and security to every stakeholder to work actively with futuristic approaches like the solution of the problems.

10.6 CONCLUSION

Learning provides the sense and wisdom to discover the truth in any society, whereas the Industry 5.0 type system provides the tools and techniques to resolve the issues of human beings at a small or large scale. Industry 5.0 is a system of the shifting of paradigm to involve smart technology, i.e., robots, to help human beings to increase and enhance the production of goods and services, because most of the countries of the world are facing a shortage of skilled workers, teachers, engineers, researchers, etc. in this era of Industry 5.0. Therefore, to meet future changes, a new concept has emerged to keep robots working in different industries so as to help humans complete their needs and requirements for goods and services. For the accomplishment of the needs of the society, highly skilled and knowledgeable human beings are needed who get quality education and then work efficiently to produce the required things and systems for the world.

Learners in the era of Industry 5.0 must be elegant and sound in technology (Nayyar and Kumar 2020) to bring advancement through their research, which can be possible only through quality education and educational institutions that can digitalize everything like learning, learning environment, research, etc. with the help of internet and ICTs. The future of quality of education is directly connected with the efforts of educational leaders and managers to maintain the things and day-to-day matters and routines in this technologically advanced era, i.e., the maintenance and usage of the IoT, and CPS. IoT and CPS support industrial applications and automation incredibly to bring change for the betterment of the people and industry in the era of Industry 5.0 to achieve the MDGs.

REFERENCES

Acemoglu, D., and Linn, J. 2003. Market size in innovation: Theory and evidence from the pharmaceutical industry. In NBER Working Paper Series. NBER, Cambridge, MA. https://doi.org/10.3386/w10038

Ahmada, H. A., and Fakhr, D. R. 2022. Relationship between organizational structure and employee behavior. *Journal of Positive School Psychology*, 6(8), 1959–1963.

Alfadala, A., Morel, R. P., and Spillane, J. P. 2021. *Multilevel distributed leadership. In Future Alternatives for Educational Leadership*, (pp. 79–92). Routledge. https://doi.org/10.4324/9781003131496-9

Al-Hosni, A. K., and Al-Ali, S. F. 2022. MOOCs in Omani higher education institutions: Use and popularity. *Education, Language and Sociology Research*, 3(1), p1. https://doi.org/10.22158/elsr.v3n1p1

Al-Sharhan, A., and Al-Hunaiyyan, S., Al-Sharhan, A. 2016. *Towards an effective integrated E-learning system: Implementation, quality assurance and competency models. In Seventh International Conference on Digital Information Management (ICDIM 2012)*. IEEE. https://doi.org/10.1109/icdim.2012.6360 142

Avota, B. 2018. Assessment of distance learning quality criteria for continuing education of medical practitioners. *SOCIETY. INTEGRATION. EDUCATION. Proceedings of the International Scientific Conference*, 5, 15–24. https://doi.org/10.17770/sie2018vol1.3089

Bafadal, I., and Arifin, I. 2020. The use of conventional communication technology as effective principal leadership strategy in strengthening the role of multi-stakeholders forum for school quality improvement. In *1st International Conference on Information Technology and Education (ICITE 2020)* (pp. 645–650). Atlantis Press.

Baig, S. A. 2023. 11 Quality control and quality assurance. In *Textile Engineering: An Introduction* (Vol. 201; pp. 201–216). De Gruyter Oldenbourg.

Bates, M. S., Phalen, L., and Moran, C. 2016. Online professional development. *Phi Delta Kappan*, 97(5), 70–73. https://doi.org/10.1177/0031721716629662

Batsila, A. 2020. Inter-learner communication and collaborative learning as quality criteria of distance vocational education and training. *European Journal of Open, Distance and E-Learning*, 22(2), 98–112. https://doi.org/10.2478/eurodl-2019-0013

Belawati, T., and Zuhairi, A. 2007. The practice of a quality assurance system in open and distance learning: A case study at Universitas Terbuka Indonesia (The Indonesia Open University). *The International Review of Research in Open and Distributed Learning*, 8(1). https://doi.org/10.19173/irrodl.v8i1.340

Biggs, J., Tang, C., and Kennedy, G. 2022. *Ebook: Teaching for Quality Learning at University 5e*. McGraw-Hill Education (UK).

Breakspear, S. 2017. Embracing Agile Leadership for Learning: How leaders can create impact despite growing complexity. *Australian Educational Leader*, 39(3), 68–71.

Budiharso, T., and Tarman, B. 2020. Improving quality education through better working conditions of academic institutes. *Journal of Ethnic and Cultural Studies*, 7(1), 99–115.

Cannadine, D. 2003. The present and the past in the English Industrial Revolution, 1880–1980. In The Industrial Revolution and Work in Nineteenth-Century Europe (pp. 2–26). Routledge. https://doi.org/10.4324/9780203415917-3

Carayannis, E. G., and Morawska-Jancelewicz, J. 2022. The futures of Europe: Society 5.0 and Industry 5.0 as driving forces of future universities. *Journal of the Knowledge Economy*, *13*(4), 3445–3471.

Childress, D., Chimier, C., Jones, C., Page, E., and Tournier, B. 2020. *Change Agents: Emerging Evidence on Instructional Leadership at the Middle Tier*. Education Development Trust.

Chinooneka, T. I. 2020. School Heads' leadership practices in enhancing quality education: Perspectives from six Rural Day Secondary Schools of Masvingo District in Zimbabwe (Doctoral dissertation).

Colangelo, E., and Bauernhansl, T. 2016. Usage of analytical services in industry today and tomorrow. *Procedia CIRP*, *57*, 276–280. https://doi.org/10.1016/j.procir.2016.11.048

Concepcion, R., Ramirez, T. J., Alejandrino, J., Janairo, A. G., Baun, J. J., Francisco, K., and Izzo, L. G. 2022. A look at the near future: Industry 5.0 boosts the potential of sustainable space agriculture. In *2022 IEEE 14th International Conference on Humanoid, Nanotechnology, Information Technology, Communication and Control, Environment, and Management (HNICEM)* (pp. 1–6). IEEE.

Crowley, K. 2014. *Child Development: A Practical Introduction*. SAGE.

Curtis, D. J., Howden, M., Curtis, F., McColm, I., Scrine, J., Blomfield, T., Reeve, I., and Ryan, T. 2013. Drama and environment: Joining forces to engage children and young people in environmental education. *Australian Journal of Environmental Education*, *29*(2), 182–201. https://doi.org/10.1017/aee.2014.5

Day, C., Gu, Q., and Sammons, P. 2016. The impact of leadership on student outcomes. *Educational Administration Quarterly*, *52*(2), 221–258. https://doi.org/10.1177/0013161x15616863

Dello-Iacovo, B. 2009. Curriculum reform and "quality education" in China: An overview. *International Journal of Educational Development*, *29*(3), 241–249.

Dewey, J. 1916. *Democracy and Education: An Introduction to the Philosophy of Education*. Create Space Independent Publishing Platform.

Dewey, J. 1938. What is social study? *The Modern Language Journal*, *23*(3), 219. https://doi.org/10.2307/317271

Diamandis, P. H., and Kotler, S. 2020. *The Future Is Fster Than You Think: How Converging Technologies Are Transforming Business, Industries, and Our Lives*. Simon & Schuster.

Didou, A. S. 2022. Quality assurance for technological institutes in Mexico: Competition and inequity. *Quality Assurance in Education*, *30*(3), 336–351.

Drexler, N., and Lapré, V. B. 2019. For better or for worse: Shaping the hospitality industry through robotics and artificial intelligence. *Research in Hospitality Management*, *9*(2), 117–120.

Dube-Xaba, Z., and Makae, M. P. 2022. Assuring quality of school-based assessment: The perspectives of heads of department in moderating tourism tasks. *Quality Assurance in Education*, *30*(1), 87–101.

Ellis, J. M. 2021. *The Breakdown of Higher Education: How It Happened, the Damage It Does, and What Can Be Done.* Encounter Books.

Eshach, H. 2007. Bridging in-school and out-of-school learning: Formal, non-formal, and informal education. *Journal of Science Education and Technology, 16*(2), 171–190. https://doi.org/10.1007/s10956-006-9027-1

Ezzani, M. D., Brooks, M. C., Yang, L., and Bloom, A. 2021. Islamic school leadership and social justice: An international review of the literature. *International Journal of Leadership in Education, 26*(10), 1–33.

Freire, P. 2018. *Pedagogy of the Oppressed: 50th Anniversary Edition.* Bloomsbury Publishing, USA.

Gates, S. M., Baird, M. D., Master, B. K., and Chavez-Herrerias, E. R. 2019. *Principal Pipelines: A Feasible, Affordable, and Effective Way for Districts to Improve Schools.* Research Report RR-2666-WF. RAND Corporation, Santa Monica, CA.

George, A. S., and George, A. H. 2020. Industrial Revolution 5.0: The transformation of the modern manufacturing process to enable man and machine to work hand in hand. *Journal of Seybold Report,* ISSN No. 1533-9211.

Giesbers, S. A., Tournier, T., Hendriks, L., Hastings, R. P., Jahoda, A., and Embregts, P. J. 2019. Measuring emotional support in family networks: Adapting the Family Network Method for individuals with a mild intellectual disability. *Journal of Applied Research in Intellectual Disabilities, 32*(1), 94–105.

Gray, J. 2018. Leadership-focused coaching: A research-based approach for supporting aspiring leaders. *International Journal of Educational Leadership Preparation, 13*(1), 100–120.

Halpern, D. F. 2010. *Undergraduate Education in Psychology: A Blueprint for the Future of the Discipline.* American Psychological Association.

Halpin, P. D., and Halpin, D. 2002. *Hope and Education.* Routledge. https://doi.org/10.4324/9780203468012

Hermann, M., Pentek, T., and Otto, B. 2016. Design principles for Industrie 4.0 scenarios. In 2016 49th Hawaii International Conference on System Sciences (HICSS), Koloa, HI, USA. https://doi.org/10.1109/hicss.2016.488

Holsombach-Ebner, C. 2013. Quality assurance in large scale online course production. *Online Journal of Distance Learning Administration, 16*(2).

Hooks, B. 2003. *Teaching Community: A Pedagogy of Hope.* Psychology Press.

Hooks, B. 2013. Teaching Community: A Pedagogy of Hope. Routledge. https://doi.org/10.1016/j.ifacol.2015.06.143

Hou, Angela Yung Chi, Robert Morse, and Chung-Lin Chiang. 2012. An analysis of mobility in global rankings: making institutional strategic plans and positioning for building world-class universities. *Higher Education Research & Development, 31*(6), 841–857.

Humayun, M. 2021. Industrial Revolution 5.0 and the role of cutting edge technologies. *International Journal of Advanced Computer Science and Applications, 12*(12).

Jensen, B., Downing, P., and Clark, A. 2017. *Preparing to Lead: Lessons in Principal Development from High-Performing Education Systems.* National Center on Education and the Economy.

Jones, E. 2014. *Quality Management for Organizations Using Lean Six Sigma Techniques.* CRC Press.

Kesler, A., Shamir-Inbal, T., and Blau, I. (2022). Active learning by visual programming: Pedagogical perspectives of instructivist and constructivist code teachers and their implications on actual teaching strategies and students' programming artifacts. *Journal of Educational Computing Research*, 60(1), 28–55.

Kishabale, B., and Hassan, S. 2018. A predictive study on instructional design quality, learner satisfaction and continuance learning intention with E-learning courses: Data screening and preliminary analysis. *Interdisciplinary Journal of Education*, 1(2), 122–137. https://doi.org/10.53449/ije.v1i2.59

Kumar, A., and Manivannan, S. 2023. *Quality Education*. NCSDSGE–2023, 124. PG & Research Department of Commerce, Coimbatore Institute of Information Technology, Coimbatore.

Lampropoulos, G., Siakas, K., and Anastasiadis, T. 2018. Internet of Things (IoT) in industry: Contemporary application domains, innovative technologies and intelligent manufacturing. *International Journal of Advances in Scientific Research and Engineering*, 4(10), 109–118.

Lampropoulos, G., Saikas, K. V., and Anastasiadis, T. 2019. Internet of Things in the context of industry 4.0: An overview. *International Journal of Entrepreneurial Knowledge*, 7(1), 4–19.

Lane, A. 2012. A review of the role of national policy and institutional mission in European distance teaching universities with respect to widening participation in higher education study through open educational resources. *Distance Education*, 33(2), 135–150. https://doi.org/10.1080/01587919.2012.692067

Lasi, H., Fettke, P., Kemper, H., Feld, T., and Hoffmann, M. 2014. Industrie 4.0. *WIRTSCHAFTSINFORMATIK*, 56(4), 261–264. https://doi.org/10.1007/s11576-014-0424-4

Leaver, C., Lemos, R. F., and Scur, D. 2019. Measuring and explaining management in schools: New approaches using public data. In *World Bank Policy Research Working Paper*. https://doi.org/10.1596/1813-9450-9053.

Lee, R. S. 2020. Smart education. In *Artificial Intelligence in Daily Life* (pp. 301–320). Springer. https://doi.org/10.1007/978-981-15-7695-9_11

Lee, S. M., and Trimi, S. 2018. Innovation for creating a smart future. *Journal of Innovation & Knowledge*, 3(1), 1–8.

Li, R. C., Cheung, S. K., Ng, P. H., Wong, L., and Wang, F. L. 2022. Blended learning: Engaging students in the new normal era. In 15th International Conference, ICBL 2022, Hong Kong, China, July 2022, proceedings. Springer Nature.

Lieberman, M. D. 2015. *Social: Why Our Brains Are Wired to Connect*. OUP, Oxford.

Litvaj, I., Drbúl, M., and Bůžek, M. 2023. Sustainability in small and medium enterprises, sustainable development in the Slovak Republic, and sustainability and quality management in small and medium enterprises. *Sustainability*, 15(3), 2039.

Longo, F., Padovano, A., Gazzaneo, L., Frangella, J., and Diaz, R. 2021. Human factors, ergonomics and Industry 4.0 in the oil & gas industry: A bibliometric analysis. *Procedia Computer Science*, 180, 1049–1058. https://doi.org/10.1016/j.procs.2021.01.350

Maddikunta, P. K. R., Pham, Q. V., Prabadevi, B., Deepa, N., Dev, K., Gadekallu, T. R., and Liyanage, M. 2022. Industry 5.0: A survey on enabling technologies and potential applications. *Journal of Industrial Information Integration*, 26, 100257.

Mahdiuon, R., Masoumi, D., and Farasatkhah, M. 2017. Quality improvement in virtual higher education: A grounded theory approach. *Turkish Online Journal of Distance Education*, 18(1), 111–111. https://doi.org/10.17718/tojde.285720

Majid, M. I., Darmawan, C. K., Majid, S. A., and Yulianto, Y. 2019. Anticipating the entry of Industry 5.0 in transportation sector. *Advances in Transportation and Logistics Research*, 2, 103–115.

Marciniak, R. 2018. Quality assurance for online higher education programmes: Design and validation of an integrative assessment model applicable to Spanish universities. *The International Review of Research in Open and Distributed Learning*, 19(2). https://doi.org/10.19173/irrodl.v19i2.3443

Marín, I. V., Zawacki-Richter, O., and Bedenlier, S. 2020. Open educational resources in German higher education – an international perspective. *EDEN Conference Proceedings* (1), 85–94. https://doi.org/10.38069/edenconf-2020-rw-0010

Marr, B. 2015. *Big Data: Using SMART Big Data, Analytics and Metrics to Make Better Decisions and Improve Performance*. John Wiley & Sons.

Miller, G., and Miller, W. 2000. A telecommunications network for distance education learning: If it's built, will agriculture teachers use it? *Journal of Agricultural Education*, 41(1), 79–87. https://doi.org/10.5032/jae.2000.01079

Mitra, A. 2016. *Fundamentals of Quality Control and Improvement*. John Wiley & Sons.

Mohammadian, H. D. 2020. IoT-Education technologies as solutions towards SMEs' educational challenges and I4.0 readiness. In *2020 IEEE Global Engineering Education Conference (EDUCON)* (pp. 1674–1683). IEEE.

Mourato, Joaquim, Maria Teresa Patrício, Luís Loures, and Helena Morgado. 2021. Strategic priorities of Portuguese higher education institutions. *Studies in Higher Education*, 46(2), 215–227.

Mowen, T. J. 2013. Parental involvement in school and the role of school security measures. *Education and Urban Society*, 47(7), 830–848. https://doi.org/10.1177/0013124513508581

Msuya, L. C., and Mwila, P. M. 2023. The effect of heads of schools' supervisory practices on learning achievement in public secondary schools in Ubungo municipality, Tanzania. *Journal of Educational and Management Studies*, 13(1), 1–42.

Mukhopadhyay, M., Pal, S., Nayyar, A., Pramanik, P. K. D., Dasgupta, N., and Choudhury, P. 2020. Facial emotion detection to assess Learner's State of mind in an online learning system. In *Proceedings of the 2020 5th International Conference on Intelligent Information Technology* (pp. 107–115). Association for Computing Machinery.

Musarat, M. A., Irfan, M., Alaloul, W. S., Maqsoom, A., and Ghufran, M. 2023. A Review on the way forward in construction through Industrial Revolution 5.0. *Sustainability*, 15(18), 13862.

Nayyar, A., and Kumar, A. (Eds.). 2020. *A Roadmap to Industry 4.0: Smart Production, Sharp Business and Sustainable Development* (pp. 1–21). Springer, Berlin.

Nazar, R., Chaudhry, I. S., Ali, S., and Faheem, M. 2018. Role of quality education for sustainable development goals (SDGS). *International Journal of Social Sciences*, 4(2), 486–501.

Nsiah, G. K., and Oti-Boadi, M. 2015. Undefined. *Creative Education*, 06(08), 707–710. https://doi.org/10.4236/ce.2015.68072

Nussbaum, M. C. 2006. Education and democratic citizenship: Capabilities and quality education. *Journal of Human Development*, 7(3), 385–395.

Onwujekwe, O., Mbachu, C., Onyebueke, V., Ogbozor, P., Arize, I., Okeke, C., and Ensor, T. 2022. Stakeholders' perspectives and willingness to institutionalize linkages between the formal health system and informal healthcare providers in urban slums in southeast, Nigeria. *BMC Health Services Research*, 22(1), 1–14.

Pandey, U., and Indrakanti, V. 2017. *Open and Distance Learning Initiatives for Sustainable Development*. IGI Global.

Parkin, D. 2022. Programme leaders as educational and academic leaders. In Supporting Course and Programme Leaders in Higher Education (pp. 95–108). Routledge. https://doi.org/10.4324/9781003127413-18

Phelps, P. H. 2008. Helping teachers become leaders. *The Clearing House: A Journal of Educational Strategies, Issues and Ideas*, 81(3), 119–122.

Pramana, C., Chamidah, D., Suyatno, S., Renadi, F., and Syaharuddin, S. 2021. Strategies to improved education quality in Indonesia: A review. *Turkish Online Journal of Qualitative Inquiry*, 12(3).

Purnamasari, F., Nanda, H. I., Anugrahani, I. S., Muqorrobin, M. M., and Juliardi, D. 2019. The late preparation of IR 4.0 and Society 5.0: Portrays on the accounting students' concerns. *South East Asian Journal of Contemporary Business, Economics and Law*, 19(5), 212–217.

Rosak-Szyrocka, J., Żywiołek, J., and Shahbaz, M. (Eds.). 2023. *Quality Management, Value Creation, and the Digital Economy*. Taylor & Francis.

Sahito, Z., and Chachar, G. B. 2021. COVID-19 and the educational leadership & management. In Akkaya, B., Jermsittiparsert, K., Malik, M. A., & Kocyigit, Y. (Eds.), *Emerging Trends and Strategies for Industry 4.0: During and Beyond COVID-19* (pp. 117–128). Sciendo Publishers. https://doi.org/10.2478/9788366675391-003

Salendab, F. A., and Dapitan, Y. C. 2021. Performance of private higher education institutions and the school heads' supervision in South Central Mindanao. *Psychology and Education*, 58(3), 3980–3997.

Scheunemann, S., Brandao, A., and Brauner, D. 2018. Towards defining quality criteria for digital educational resources in distance learning. In 2018 IEEE World Engineering Education Conference (EDUNINE). IEEE. https://doi.org/10.1109/edunine.2018.8450968

Shakeshaft, C., Brown, G., Irby, B., Grogan, M., and Ballenger, J. 2014. Increasing gender equity in educational leadership. In *Handbook for Achieving Gender Equity through Education* (pp. 133–160). Routledge.

Show, P. L., Chew, K. W., and Ling, T. C. (Eds.). 2021. *The Prospect of Industry 5.0 in Biomanufacturing*. CRC Press.

Smith, C. D. 2008. Secondary special education. In Encyclopedia of Special Education. Wiley. https://doi.org/10.1002/9780470373699.speced1865

Stambekova, A., Zhakipbekova, S., Tussubekova, K., Mazhinov, B., Shmidt, M., and Rymhanova, A. 2022. Education for the disabled in accordance with the quality of inclusive education in the distance education process. *World Journal on Educational Technology: Current Issues*, 14(1), 316–328. https://doi.org/10.18844/wjet.v14i1.6760

Strauss, C. 1984. Beyond "Formal" versus "Informal" Education: Uses of psychological theory in anthropological research. *Ethos*, 12(3), 195–222. https://doi.org/10.1525/eth.1984.12.3.02a00010

Strickland, H. P., and Kaylor, S. K. 2016. Bringing your a-game: Educational gaming for student success. *Nurse Education Today*, 40, 101–103. https://doi.org/10.1016/j.nedt.2016.02.014

Taftaf, R., and Williams, C. 2019. Supporting refugee distance education: A review of the literature. *American Journal of Distance Education*, 34(1), 5–18. https://doi.org/10.1080/08923647.2020.1691411

Taj, I., and Zaman, N. 2022. Towards Industrial Revolution 5.0 and explainable artificial intelligence: Challenges and opportunities. *International Journal of Computing and Digital Systems*, 12(1), 295–320.

Tavares, M. C., Azevedo, G., and Marques, R. P. 2022. The challenges and opportunities of era 5.0 for a more humanistic and sustainable society – a literature review. *Societies*, 12(6), 149.

Tudor, S. L. 2013. Formal–non-formal–informal in education. *Procedia – Social and Behavioral Sciences*, 76, 821–826. https://doi.org/10.1016/j.sbspro.2013.04.213

Ullah, S. M., Bodrogi, A., Cristea, O., Johnson, M., & McAlister, V. C. 2012. Learning surgically oriented anatomy in a student-run extracurricular club: An education through recreation initiative. *Anatomical Sciences Education*, 5(3), 165–170. https://doi.org/10.1002/ase.1273

Vaidya, S., Ambad, P., and Bhosle, S. 2018. Industry 4.0: A glimpse. *Procedia Manufacturing*, 20, 233–238. https://doi.org/10.1016/j.promfg.2018.02.034

Visvizi, A., Daniela, L., and Chen, C. W. 2020. Beyond the ICT and sustainability hypes: A case for quality education. *Computers in Human Behavior*, 107, 106304.

Vlachopoulos, D., and Makri, A. 2019. Online communication and interaction in distance higher education: A framework study of good practice. *International Review of Education*, 65(4), 605–632. https://doi.org/10.1007/s11159-019-09792-3

Von Tunzelmann, N. 2003. Historical coevolution of governance and technology in the industrial revolutions. *Structural Change and Economic Dynamics*, 14(4), 365–384. https://doi.org/10.1016/s0954-349x(03)00029-8

Warnock, M. 1986. Another ten years in education. *Journal of the Royal Society of Medicine*, 79(4), 194–199. https://doi.org/10.1177/014107688607900403

White, M. A., and McCallum, F. 2020. Critical perspectives on teachers and teaching: An appreciative examination. In *Critical Perspectives on Teaching, Learning and Leadership* (pp. 1–15). Springer, Singapore.

Yin, Yong, Kathryn E. Stecke, and Dongni Li. 2018. The evolution of production systems from Industry 2.0 through Industry 4.0. *International Journal of Production Research*, 56(1–2), 848–861.

Yun, Jinhyo Joseph, MinHwa Lee, KyungBae Park, and Xiaofei Zhao. 2019. Open innovation and serial entrepreneurs. *Sustainability*, 11(18), 5055.

Zawacki-Richter, Olaf, and Adnan Qayyum. 2019. *Open and Distance Education in Asia, Africa and the Middle East: National Perspectives in a Digital Age.* Springer Nature.

Chapter 11

Traditional pedagogy to Technetronic Education

A paradigm shift in Industry 5.0

Aparna Vajpayee

11.1 INTRODUCTION

11.1.1 Conceptual background

The advent of Industry 5.0 brings forth a revolutionary shift not only in manufacturing and production but also in education. As we stand on the precipice of a new industrial era, the education sector undergoes a profound transformation, adapting to the changing needs of society and industry alike (Al-Rahmi, Othman, and Musa, 2014; Anderson and Garrison, 1995). In this dynamic landscape, the fusion of traditional teaching methodologies with cutting-edge technological advancements is paramount to preparing the workforce of tomorrow (Lauricella and Kay, 2013).

Industry 5.0 emphasizes the integration of human cognitive capabilities with advanced technologies, creating a symbiotic relationship where humans and machines collaborate for enhanced productivity and innovation (Vajpayee and Ramchandran,, 2019). This paradigm shift extends to education, giving rise to what we can term as Technetronic Education (TE), a concept that aligns perfectly with the principles of Industry 5.0. TE encapsulates the convergence of traditional pedagogy with state-of-the-art technology to create a holistic learning experience. It leverages several key aspects:

i. **Technological Integration:** TE seamlessly incorporates a plethora of technologies such as artificial intelligence (AI), augmented reality, virtual reality, the Internet of Things, and advanced data analytics into the educational ecosystem. These technologies empower educators and learners alike by providing tools for immersive and interactive learning and flexible pedagogy for teaching (Gordon, 2014)

ii. **Personalized Learning:** In the Industry 5.0 era, education is tailored to individual needs. TE harnesses the power of data-driven insights to create personalized learning pathways. This ensures that each learner

DOI: 10.1201/9781032677040-11

      receives education that aligns with their strengths, weaknesses, and interests.

iii. **Collaborative Learning Environments:** Industry 5.0 thrives on collaboration, and so does TE. It fosters collaborative learning environments where students engage in group projects, real-world problem-solving, and interactive discussions, mirroring the collaborative nature of modern workplaces (Arasaratnam and Northcote, 2017).

iv. **Lifelong Learning:** Industry 5.0 demands that individuals engage in lifelong learning to adapt to the ever-evolving technological landscape. TE embraces this philosophy by providing accessible, continuous learning opportunities beyond traditional classrooms (Criollo-C, Luján- Mora, and Jaramillo-Alcázar, 2018).

v. **Human–Machine Synergy:** TE prepares students to work alongside advanced technologies, nurturing skills such as critical thinking, creativity, and adaptability. It emphasizes that humans are not replaced by machines but work in tandem with them (Anderson and Garrison, 1995).

While the advent of TE and Industry 5.0 heralds a promising future for education and the workforce, it also introduces certain challenges. The potential drawbacks include an overreliance on technology, reduced face-to-face interactions, and concerns regarding data privacy and security. It is imperative to strike a balance between the digital and human aspects of education to harness the full potential of TE within Industry 5.0.

Research conducted by Cochran and Borbieva (2023) found that the integration of technology in education, as exemplified by Industry 5.0 and Technology-Enhanced Education, can significantly enhance learning outcomes and provide students with valuable skills for the future workforce. However, their study also pointed out that there are substantial concerns associated with this shift. Smith et al. emphasized that an overreliance on technology might lead to disengagement, reduced face-to-face interactions, and potential privacy and security risks, which need to be carefully addressed in the evolving educational landscape (Sheokand, Vajpayee and Sanghani, 2023). This underscores the importance of finding a delicate balance between the digital and human aspects of education to realize the full potential of TE within Industry 5.0.

This research delves into the heart of this transformative process, exploring the advantages, challenges, and potential pitfalls of TE within the context of Industry 5.0. It aims to provide valuable insights for educators, policymakers, and industry leaders as they navigate this exciting and dynamic educational landscape. As Industry 5.0 unfolds, education stands at the forefront of change, ready to empower individuals with the knowledge and skills needed to thrive in a technology-driven world (Padmanabhan, 2020).

COVID-19 brought about the community being restricted (Vajpayee, Devnani, and Sanghani, 2023) to online technological educational environments in lieu of traditional ways of teaching for a long time; it represented a paradigm change in education. As a result, the adoption ability of conventional approaches to teaching versus cutting-edge tech educational methods has been discussed. When compared to conventional traditional classroom modality, the notion of quality educational instruction via the web possesses its own benefits and challenges that require to be embraced and optimized. In contrary to the above argument that conventional classes are limited in their available modalities, inflexible and unworkable still have a certain unique unreplaceable impotence. The widespread implementation of advice technologies in education also comes with hazards for a big populace. Academicians and academic organizations need to re-evaluate the manner in which to implement their curricula by employing pedagogical means and revolutionary technological methods.

Furthermore, there are an array of drawbacks associated with techno-dominant education, such as fewer opportunities for social interaction, loneliness, a lack of practical experience, a lack of community integration, a disproportionate emphasis on the individualization of everything, a lack of social skills, stress anxiety, and depression brought on by isolation. The difficulties of screen addictions need to be addressed along with other significant dangers like a lack of social adjustment, impatience, a lack of group thinking, and nomophobia with techno-dominant education (Devnani et al, 2022; Vajpayee, Devnani, and Sanghani, 2023).

11.1.2 Technetronic Education

Technetronic Education (TE) is a term coined by American sociologist and futurist Alvin Toffler in his book *The Third Wave* published in 1980. The concept of TE explores the idea of utilizing advanced technology and electronic systems to enhance and transform the process of education. Toffler envisioned TE as a shift from traditional classroom-based learning to a more technologically mediated and interactive approach. He believed that emerging technologies would revolutionize education by providing new tools, resources, and methods of instruction. The key aspects of TE can include the following:

i. **Technological Integration:** This refers to the incorporation of numerous technologies into the educational process, including computers, multimedia, the internet, virtual reality, and AI. These technologies may be used to distribute material, promote individualized learning pathways, and enable interactive learning experiences.

ii. **Information Access:** The quick and extensive information access made possible by technology is emphasized in technological education. Through digital libraries, instructional databases, and online platforms, students have access to a tremendous amount of information and materials.

iii. **Individualized Learning:** Because of technology, learning can be flexible and customized to the interests and needs of each learner. Data on student performance may be analyzed by intelligent algorithms, which can then offer personalized learning routes and feedback.

iv. **Collaborative and Interactive Learning:** Technology-enabled learning settings are encouraged by technological education. Students can participate in online debates, team projects, and simulations to improve their comprehension and critical thinking abilities.

v.   **Lifelong Learning:** In a world that is changing quickly, technological education recognizes the value of ongoing education. Through digital platforms, online courses, and distance learning options, it advocates the concept that education should go beyond traditional schools and facilitate lifelong learning.

vi. **Collaborative and Interactive Learning:** Technology-enabled learning settings are encouraged by technological education. Students can participate in online debates, team projects, and simulations to improve their comprehension and critical thinking abilities (Anderson, and Garrison, 1995).

vii. **Lifelong Learning:** In a world that is changing quickly, technological education recognizes the value of ongoing education. Through digital platforms, online courses, and distance learning options, it advocates the concept that education should go beyond traditional schools and facilitate lifelong learning.

11.1.3 Traditional education

Traditional education plays an important role in strengthening the cultural values and traditions of a society. It also helps to preserve the knowledge, language, and values passed down from generations. Traditional education often focuses on developing skills for use in everyday life, such as reading, writing, arithmetic, agricultural, and spatial techniques along with shaping cognition (Vajpayee, Mishra, and Dasen, 2008). By providing these skills, individuals are able to successfully complete tasks that are necessary for daily living (Todd, Towne, and Clarke, 2023).

Traditional education allows people to gain knowledge and understanding of the principles of their culture, making them more likely to carry on the values and beliefs of their particular society. It also helps students to understand and appreciate the history of their community and culture (Mishra and Vajpayee, 2004). Through traditional education, students become more aware of the traditional stories, beliefs, norms, and widely accepted behavioral patterns that shape their culture. This helps them to build relationships with those of their own communities and retain a sense of belonging to their homeland. Traditional education also provides students the opportunity to develop creativity by testing their innovative ideas and trial-and-error methods outside the traditional classroom setting. Through traditional education, students gain the skills and knowledge to become successful in an economic and political sense and aligned with a sense of belongingness to their culture (Santosh and Panda, 2016) along with in-depth understanding of cultural and moral orientation (Kantharia,, Vajpayee, and Sanghani, 2023; Vadher, Vajpayee, and Sanghani, 2023).

Mental health is an essential factor in the well-being of individuals and their communities. Traditional education has the potential to positively affect the mental well-being of students by providing the environment to discuss personal issues and concerns. Through traditional education, students are provided with the platform to engage in dialogue about their own mental health challenges and experiences without judgment (Goel et al, 2019). Traditional education also provides opportunities to learn more about mental health and develop a better understanding of mental health issues (Zheng, Bao, and Wang, 2022). Traditional educational systems provide integrated mental health services and support systems which provide students with guidance and resources necessary for adapting to life's challenges and living happily and productively with the longevity of life (Vajpayee and Sanghani, 2022).

Understanding the impact of the evolving educational paradigm within the context of Industry 5.0 is of paramount importance. This research aims to shed light on the enduring relevance and adaptability of the new educational system as it aligns with the broader goals and transformations of Industry 5.0. While past studies have meticulously dissected individual components of technology in higher education, this research takes a holistic approach, exploring how these components synergize and interact within the ever-evolving educational landscape.

In the era of Industry 5.0, where human–machine collaboration is central, the study delves into the comprehensive advantages and disadvantages of TE in the lives of students and educators. It seeks to delineate the nuanced interplay of these advantages and drawbacks while drawing insightful comparisons with traditional educational methods (Edmonds, 2004; Goel et al, 2019). Moreover, it identifies specific domains within which traditional education maintains its unique significance, offering a balanced perspective on preserving time-tested teaching practices.

11.2 OBJECTIVE OF THE RESEARCH WITHIN THE FRAMEWORK OF INDUSTRY 5.0

i. **Advantages of TE in Industry 5.0 Education:** This research strives to uncover the multifaceted advantages of TE within the context of Industry 5.0, demonstrating how it equips learners with the skills and knowledge needed to thrive in technologically driven industries.

ii. **TE as a Teaching Aid in Industry 5.0:** The study explores the versatile role of TE as a teaching aid, examining its impact on various aspects of teaching, classroom management, and individualized learning in the Industry 5.0 environment.

iii. **Impact of TE on Personalized Learning and Teaching in Industry 5.0:** It investigates how TE fosters personalized learning experiences and enables educators to tailor their teaching methods to meet diverse needs of students, aligning with the principles of Industry 5.0's human–technology synergy (Grant and Basye, 2014).

iv. **Online Learning and Interpersonal Dynamics in Industry 5.0:** The research delves into the significance of TE in fostering effective online communication, preserving teacher–student relationships, and addressing mental health concerns such as anxiety, depression, and loneliness in the digital age.

v. **Long-term Perspectives of TE in Industry 5.0:** It explores the potential for TE to complement, rather than replace, teachers in Industry 5.0, emphasizing the enduring importance of human guidance and mentorship (Haleem et al, 2019).

vi. **TE Across Domains in Industry 5.0:** This study investigates how TE extends its influence across diverse domains, showcasing its adaptability and relevance in various fields and industries associated with Industry 5.0.

vii. **Limitations of TE in Personalized Industry 5.0 Education:** The research critically assesses the limitations of TE, particularly in the context of delivering domain-specific and highly personalized education within the dynamic landscape of Industry 5.0.

viii. **Indirect Impacts on Education within Industry 5.0:** Beyond the immediate classroom, the study identifies the indirect impacts of TE on the broader educational ecosystem, including implications for curriculum design, institutional structures, and the professional lives of educators.

This research is a comprehensive exploration of the evolving educational landscape in the age of Industry 5.0, aiming to provide valuable insights into the intricate dynamics between technology-driven education and traditional methods. It seeks to inform educators, policymakers, and industry leaders as they navigate the transformative journey of education within Industry 5.0's innovative and collaborative framework.

11.3 METHODOLOGY

In our pursuit of understanding the comparative advantages of traditional education versus TE within the Industry 5.0 landscape, we employed a robust methodology that resonates with the demands of this transformative era.

Sampling Approach: To gauge the efficacy and prevalence of TE in educational practices, we adopted a probability sampling approach. Specifically, we collected primary data from a pool of 300 research participants who had experienced extended periods of online learning within the context of Industry 5.0.

Participant Selection Criteria: In alignment with the dynamic requirements of Industry 5.0, we specifically targeted academicians who had been actively incorporating technology into their teaching methodologies for a duration exceeding 6 months. Our sample pool consisted of professors spanning various age groups and academic specialties, including engineering, medical care, nursing, management, and liberal arts. This diverse selection ensured that we comprehensively understood the domain-specific needs and nuances of technology integration within Industry 5.0 education.

11.3.1 Data collection methods

Google Questionnaire: We employed a Google-based questionnaire to efficiently gather quantitative data on participants' experiences and perceptions of online learning and TE. This method allowed us to obtain structured and quantifiable insights into the extent of technological incorporation.

Structured Interviews: Recognizing the complexity of the Industry 5.0 educational landscape, we complemented our quantitative data collection with structured interviews. These interviews provided a qualitative dimension to our research, enabling us to delve deeper into participants' technological approaches and the nuances of their experiences.

11.3.2 Comprehensive insight into technological approach

By combining quantitative data from the questionnaire with qualitative insights from structured interviews, we aimed to gain a holistic understanding of the technological strategies and practices embraced by academicians operating within Industry 5.0. This blended approach allowed us to capture both the quantitative prevalence and the qualitative intricacies of TE in this transformative era (Nawaz and Khan, 2012).

Our research methodology was tailored to the specific requirements of Industry 5.0, where technology and education converge to shape the future of learning. By engaging with academicians who have actively embraced technology in their teaching, we sought to provide valuable insights into the evolving educational paradigms within this innovative industrial landscape (Qureshi et al, 2021).

11.4 ANALYSIS OF THE RESULTS

Table 11.1 reveals that within the context of Industry 5.0, teachers are actively and extensively embracing TE as a preferred approach in higher academics.

Table 11.1 highlights the shift toward TE, which is indicative of the transformative impact of Industry 5.0 on education. The findings underscore the profound impact of technology on the educational landscape in Industry 5.0. As industries evolve and embrace advanced technologies, education must adapt to prepare students for the demands of this innovative era. TE emerges as a pivotal tool in achieving this objective, facilitating communication, motivation, competence, and personalized learning experiences in alignment with the principles of Industry 5.0. Table 11.2 shows that the factor analysis of positive impacts associated with TE in higher education illuminates a nuanced landscape.

It reveals that TE is perceived positively for its potential to enhance academic achievement and effectiveness, providing a valuable instructional tool that fosters teachers' confidence and eases their professional pressures. However, concerns are raised about potential drawbacks, including the neglect of traditional learning resources and the stress-inducing nature of increased reliance on technology. Furthermore, Table 11.2 underscores the versatile role of TE in instructional practices, proficiency development, and creative applications. Teachers express a readiness to adapt to TE's informative capabilities, yet apprehensions persist regarding its potential impact on future employment opportunities for educators. While TE is recognized for its expansive use in fostering students' interpersonal skills and engaging them in creative activities, challenges such as time demands and potential

Table 11.1 Implementation of Technetronic Education for higher education

KMO and Bartlett's Test[a]		
Kaiser–Meyer–Olkin Measure of Sampling Adequacy.		*0.646*
Bartlett's Test of Sphericity	Approx. Chi-Square	1278.469
	df	561
	Sig.	0.000

[a] Based on correlations.

Table 11.2 Factor analysis: Positive impact of Technetronic Education for higher education

Communalities	Extraction
1. Increases academic achievement (e.g., grades).	0.839
2. Results in students neglecting important traditional learning resources (e.g., library books).	0.856
3. Results in students neglecting important traditional learning resources (e.g., library books).	0.829
4. Is effective because I believe I can implement it successfully	0.678
5. Makes classroom management more difficult.	0.546
6. Promotes the development of communication skills (e.g., writing and presentation skills).	0.656
7. Is a valuable instructional tool.	0.578
8. Makes teachers feel more competent as educators.	0.721
9. Gives teachers the opportunity to be learning facilitators instead of information providers	0.663
10. Demands that too much time be spent on technical problems	0.649
11. Is an effective tool for students of all abilities	0.835
12. Enhances my professional development	0.670
13. Eases the pressure on me as a teacher	0.622
14. Is effective if teachers participate in the selection of computer technologies to be integrated	0.708
15. Helps accommodate students' personal learning styles.	0.696
16. Motivates students to get more involved in learning activities.	0.664
17. Could reduce the number of teachers employed in the future.	0.629
18. Limits my choices of instructional materials.	0.682
19. Requires software-skills training that is too time consuming	0.502
20. Promotes the development of students' interpersonal skills (e.g., ability to relate or work with others).	0.702
21. Will increase the amount of stress and anxiety students experience	0.764
22. Requires extra time to plan learning activities.	0.712
23. Improves student learning of critical concepts and ideas.	0.671
24. Is Interactive panel your preferred teaching methodology (choose only one)	0.744
25. Please read the following descriptions of the proficiency levels (Interactive panel) a user has in relation to computer technologies. Determine the level that best describes you and circle the corresponding letter on your answer sheet.	0.521
Instructional	0.663
1. Use WebQuests in your lessons.	
2. Use tutorials for self-training.	
3. Have students use tutorials for remediation (in class).	
Organizational	0.624
1. Keep track of student grades or marks.	
2. Prepare handouts, tests/quizzes, and homework assignments for students.	
3. Create lesson plans.	

(continued)

Table 11.2 (Cont.)

Communalities	Extraction
Analytical/Programming	0.718
1. Create charts or graphs.	
2. Create a class/school website or put student work on-line.	
3. Analyze data.	
Statistics or data analysis	
Recreational	0.651
1. Have students play games (in class).	
2. Use computer time as a reward for completing class work or good behavior.	
Expansive	0.558
1. Have students conduct experiments or laboratory exercises (in class/school lab).	
2. Have students use 3-D modelling software or simulations (in class/school lab).	
Creative	0.694
1. Use drawing or paint programs.	
2. Scan pictures or images.	
Use digital video, digital cameras	
Expressive	0.652
1. Use a word processor.	
2. Maintain an online journal (diary) or discussion board.	
Evaluative	0.610
1. Test or assess student learning.	
2. Use digital portfolios.	
Informative	0.765
1. Search the internet for information for a lesson.	
2. Access CD-ROM reference material.	

disruptions to classroom management are also acknowledged. Overall, the findings indicate a dynamic interplay between the positive aspects and challenges associated with TE in the realm of higher education. Table 11.3 provides information on additional facets related to technology use in education, focusing on variables associated with expansive activities.

Improves student learning of critical concepts and ideas (Variable 23):

i. The majority of respondents (61) believe that utilizing technology improves student learning of critical concepts and ideas, as indicated by the high count for a rating of 4.0.
ii. A significant portion of respondents (28) also perceive improvement, though to a slightly lesser extent, as reflected by the count for a rating of 3.0.

Table 11.3 Additional facets related to technology use in education

Variable		Expansive 1. Have students conduct experiments or laboratory exercises (in class/school lab). 2. Have students use 3-D modelling software or simulations (in class/school lab).				Total
		1.0	3.0	4.0	5.0	
23. Improves student	2.0	1	2	0	2	5
learning of critical	3.0	5	21	0	2	28
concepts and ideas.	4.0	6	47	1	7	61
	5.0	2	4	0	2	8
Total		14	74	1	13	102

iii. There are lower counts for ratings of 2.0 and 5.0, suggesting that fewer respondents think the impact is either minimal or extremely significant.
iv. It is noteworthy that there is no response with a rating of 1.0 in this category.

Table 11.3 shows that respondents generally agree that incorporating technology, particularly through activities such as experiments, laboratory exercises, and the use of three-dimensional (3-D) modeling software or simulations, has a positive impact on improving student learning of critical concepts and ideas. The absence of responses with a rating of 1.0 indicates a consensus among respondents that these technology-related activities are not perceived as having a negative impact in this context. The data in Table 11.4 illustrates the perceived impact of TE on students' learning across two variables.

The data in Table 11.4 illustrates the perceived impact of TE on students' learning across two variables.

Improves student learning of critical concepts and ideas (Variable 23):

i. The majority of respondents (61) believe that TE significantly enhances students' understanding of critical concepts and ideas, as indicated by the high count for a rating of 4.0.
ii. A substantial portion of respondents (28) also recognize improvement, though to a slightly lesser extent, as reflected by the count for a rating of 3.0.
iii. Fewer respondents expressed concerns, with lower counts for ratings of 2.0 and 5.0.

Table 11.4 Impact of Technetronic Education in student' learning

Variable		2. Results in students neglecting important traditional learning resources (e.g., library books).				Total
		2.0	3.0	4.0	5.0	
23. Improves	2.0	1	0	1	3	5
student learning	3.0	4	12	8	4	28
of critical	4.0	12	10	31	8	61
concepts and	5.0	1	4	2	1	8
ideas.						
Total		18	26	42	16	102

Results in students neglecting important traditional learning resources (Variable 23):

i.   A significant number of respondents (42) are concerned that TE leads to the neglect of essential traditional learning resources, as evidenced by the high count for a rating of 4.0.

ii. A noteworthy but slightly smaller group of respondents (26) shares this concern but to a lesser degree, as indicated by the count for a rating of 3.0.

iii. Fewer respondents perceive neglect of traditional resources as either not a problem or a minor concern, as reflected in the lower counts for ratings of 2.0 and 5.0.

Table 11.4 shows that the majority of respondents acknowledge the positive impact of TE on students' understanding of critical concepts and ideas. However, there is a significant concern among respondents about the potential neglect of important traditional learning resources (Library and proximity learning) due to the adoption of TE. Table 11.5 outlines the impact of TE on the development of communication skills, specifically focusing on the variable "Promotes the development of communication skills (e.g., writing and presentation skills)." The impact levels are categorized from 2.0 to 5.0.

Promotes the development of communication skills (Variable 6):

i.   The table indicates that respondents recognize the role of TE in promoting the development of communication skills.

ii. A significant number of respondents (48) provided a rating of 4.0, suggesting a consensus that TE contributes positively to communication skills.

Table 11.5 Technetronic Education and critical aspects of students' learning

Variable		6. Promotes the development of communication skills (e.g., writing and presentation skills).				Total
		2.0	3.0	4.0	5.0	
23. Improves student	2.0	1	0	3	1	5
learning of critical	3.0	2	12	6	8	28
concepts and ideas.	4.0	2	9	35	15	61
	5.0	1	2	4	1	8
Total		6	23	48	25	102

Table 11.6 Technetronic Education and fear of job insecurity among teachers of higher education

*Could reduce the number of teachers employed in the future. * 5. Makes classroom management more difficult. Crosstabulation*

Variable		5. Makes classroom management more difficult.				Total
		2.0	3.0	4.0	5.0	
17. Could reduce the	2.0	21	8	5	4	38
number of teachers	3.0	6	12	6	1	25
employed in the	4.0	12	3	5	2	22
future.	5.0	8	3	3	3	17
Total		47	26	19	10	102

iii. Ratings of 3.0 and 5.0 also show substantial recognition, with counts of 23 and 25, respectively.

iv. Fewer respondents provided a rating of 2.0, indicating a smaller portion perceiving minimal impact.

Table 11.5 highlights that the majority of respondents believe that TE positively influences student learning of critical concepts and ideas, with a significant count for a rating of 4.0. There is also consensus among respondents that TE promotes the development of communication skills, particularly with a high count for a rating of 4.0. Table 11.6 presents a crosstabulation of responses related to two variables: "Could reduce the number of teachers employed in the future" and "Makes classroom management more difficult." The variables are represented by rows and columns, with impact levels ranging from 2.0 to 5.0. The "Total" row provides the sum for each column.

Could reduce the number of teachers employed in the future (Variable 17):

i.   The majority of respondents (38) express concerns that the integration of technology could potentially lead to a reduction in the number of teachers employed in the future, as indicated by the high count for a rating of 2.0.
ii. A significant number of respondents (25) also share this concern, though to a slightly lesser extent, as reflected by the count for a rating of 3.0.
iii. There are fewer respondents who believe that this impact is more significant, as evidenced by lower counts for ratings of 4.0 and 5.0.

Makes classroom management more difficult (Variable 5):

i.   The majority of respondents (47) believe that incorporating technology makes classroom management more difficult, as indicated by the high count for a rating of 2.0.
ii. A notable number of respondents (26) share this concern, though to a lesser extent, as reflected by the count for a rating of 3.0.
iii. There are fewer concerns expressed for ratings of 4.0 and 5.0.

Table 11.6 represents the findings with significant concerns that integrating technology into education could potentially reduce the number of teachers employed in the future. There is also a consensus among respondents that the use of technology makes classroom management more difficult, with a majority expressing this concern.

Table 11.7 presents a crosstabulation of responses related to three variables: "Will increase the amount of stress and anxiety students experience," "Could reduce the number of teachers employed in the future," and "Requires extra time to plan learning activities."

Will increase the amount of stress and anxiety students experience (Variable 21):

i.   A considerable number of respondents express concerns that integrating technology will increase the amount of stress and anxiety students experience, as indicated by the high count for a rating of 4.0.
ii. There are varying levels of agreement among respondents for other impact ratings (2.0, 3.0, and 5.0), with the highest count being for a rating of 5.0.

Could reduce the number of teachers employed in the future (Variable 17):

i.   Respondents show concerns that integrating technology could potentially reduce the number of teachers employed in the future. The

Table 11.7 Mental health disadvantages associated with Technetronic Education

Count: *Will increase the amount of stress and anxiety students experience * Could reduce the number of teachers employed in the future. * Requires extra time to plan learning activities. Crosstabulation.*

22. Requires extra time to plan learning activities.			17. Could reduce the number of teachers employed in the future.				Total
			2.0	3.0	4.0	5.0	
2.0	21. Will increase the amount of stress and anxiety students experience	2.0	0	1	0	0	1
		3.0	2	0	0	0	2
		4.0	7	2	1	1	11
		5.0	3	1	0	1	5
	Total		12	4	1	2	19
3.0	21. Will increase the amount of stress and anxiety students experience	2.0	1	0	0	2	3
		3.0	0	3	0	0	3
		4.0	5	3	2	2	12
		5.0	4	7	1	2	14
	Total		10	13	3	6	32
4.0	21. Will increase the amount of stress and anxiety students experience	1.0	0	0	1	0	1
		2.0	3	2	5	3	13
		3.0	1	0	1	0	2
		4.0	6	2	9	2	19
		5.0	1	4	1	2	8
	Total		11	8	17	7	43
5.0	21. Will increase the amount of stress and anxiety students experience	1.0	1		0	0	1
		3.0	0		1	0	1
		4.0	0		0	1	1
		5.0	4		0	1	5
	Total		5		1	2	8
Total	21. Will increase the amount of stress and anxiety students experience	1.0	1	0	1	0	2
		2.0	4	3	5	5	17
		3.0	3	3	2	0	8
		4.0	18	7	12	6	43
		5.0	12	12	2	6	32
	Total		38	25	22	17	102

    majority of respondents express this concern, with the highest count for a rating of 4.0.

ii. Lower counts are observed for ratings of 2.0, 3.0, and 5.0.

Requires extra time to plan learning activities (Variable 22):

i. A significant number of respondents believe that integrating technology requires extra time to plan learning activities, as indicated by the high count for a rating of 4.0.

ii. There are varying levels of agreement among respondents for other impact ratings (2.0, 3.0, and 5.0), with the highest count being for a rating of 3.0.

iii. The majority of respondents express concerns that integrating technology may lead to an increase in the amount of stress and anxiety experienced by students.

Table 11.7 shows consensus among respondents that incorporating technology could potentially reduce the number of teachers employed in the future. Additionally, respondents agree that integrating technology requires extra time to plan learning activities, with the highest count for a rating of 4.0.

11.5 COMPREHENSIVE ANALYSIS

The findings demonstrate the significant influence of technology on higher education. The majority of academics agree that technological education improves all academic results, but they also agree that conventional learning resources will be mainly ignored in the future. Implementation is simple, but planning requires more time. Teachers experience less pressure because they feel more capable and like a useful resource.

In contrast to the aforementioned arguments, they have also believed that technological education restricts their options for learning material, they often update their knowledge in light of current technological advancements, and they harbor concerns that technological advancements may result in fewer employment being available. They have acknowledged that a greater reliance on technology may cause students to experience more anxiety and stress, which is unhealthy for the student's mental health. It can enhance depression and loneliness as students will become more involved with technology and people touch will be reduced (Devnani et al, 2022).

i. **Women and Men Users of TE:** All of the participants in the study were professors of higher education, with 65.3% of the participants being men and 34.7% being women. According to their response, 55.4% of instructors agreed to use TE. The usage of technology raises students' academic success. Seventy-one percent of instructors were capable of using technological abilities in the classroom. Additionally, 71% of the instructors surveyed said they encouraged the growth of communication skills linked to writing and presenting regardless of their gender. The use of instructional technologies, according to 73% of instructors, improves educational quality and increases student competency (80% positive response by both the genders). Technology was used as a facilitator by higher education lecturers to disseminate knowledge (Rinekso and Muslim, 2020; Sabban, 2015).

ii. **TE as a Facilitator in Higher Education:** Technology helps teachers improve professionally, according to 67% of them, and it also lessens their workload and makes it easier for them to work (49%). With their reply, they have also provided a narrative. Regarding meeting pupils' individual learning styles, 63%. Although teachers were able to accommodate students' chosen learning styles, 37% of them either were not convinced by the students' preferences or were ambivalent (26.7%).

iii. **TE as a Tool to Motivate the Learner:** Seventy-five percent of teachers agree that technology may motivate students to become involved in learning activities, but they also emphasize that an ongoing spike in the usage of technology could lead to stress, anxiety, and depression among youngsters. They discuss in person how an over-reliance on technology may result in feelings of isolation and despair (Kew et al., 2018; Kärkkäinen ; Vencent-Lancrin, 2013).

iv. **TE and Creativity:** Teachers need more time to specify learning activities, according to 47% of them. In addition, technology has been useful in introducing important ideas and concepts for 65% of instructors. Sixty percent of educators believe that interactive panels are helpful tools for giving lectures. Only 41% of university professors consider themselves good lecturers when it comes to employing technology at advanced and expert levels. Only 27% of instructors are comfortable using tutorials and remediation; however, this percentage is growing.

v. **TE as a Tool of Evaluation and Recreational Activity:** TE is not a preferred way for adopting recreational activities. Only 21% of teachers utilize it to keep track of their students' grades, marks, exams, quizzes, homework assignments, and lesson plans. Fifty-six percent of instructors use technology to assist in creating graphs, charts, and analyses of data. Technology is only significant to 36% of instructors when it comes to classroom fun, and 29% of teachers solely utilize it for 3-D model software, simulations, and laboratory activities. When it comes to scanning photos and photographs, I feel the same way. Twenty-seven percent of instructors utilize digital cameras and video. The amount of people using technology as a word processor and participating in online journal discussions is similarly quite low (25%). Internet information research and compact disc (CD) access. Hence, creativity and recreational activitystill do not have reliance on TE.

vi. **TE in place of Traditional Library:** Reliance on the traditional library is high as compared to TE. Reference material on read only memory (ROM) 57% of educators feel at ease using technology. In comparison to the value of conventional learning tools, such as accessing libraries to expand one's knowledge, technological education still has a relatively restricted application (55%). This demonstrates the widespread support for using conventional libraries as learning resources among instructors. Still, most of the instructors do not rely on internet sources (Vajpayee, Sheokand, and Sanghani, 2022).

The use of technology in the classroom presents managerial challenges for many instructors (20%). Higher academic professors said that it takes up too much time, 46% of them. According to 31% of instructors, AI and smart technology may make teachers less employable in the future. Twenty-five percent of instructors said that the usage of AI had limited their options for teaching materials. Teachers need to develop their digital consumption skills, according to 43% of them.

vii. **TE and Interpersonal Communication:** Many of them have discovered that because there are no interpersonal encounters, the likelihood of developing interpersonal skills is reduced with the feeling of ostracism (Knausenberger et al., 2022) yet with traditional small size of organization with less technology usage communication and togetherness was more promising (Vajpayee and Ramchandran, 2019). Although the ratio is not extremely high, several of them have been linked to student stress and anxiety. Only 16% of those who could use technology proficiently and 29% of people who could use it advanced were in higher education, making up the remaining 45%. Technical education still requires teachers to continue to enhance these abilities in their students (Hartzell et al., 2016; Kumar, Singh, and Paddakanya, 2021).

viii. **Trust on TE:** However, it is a style of education that is frequently employed when it comes to grading pupils and creating lesson plans. Teachers still do not utilize it for making graphs, charts, and statistics in 44% of cases. Nearly 70% never rely on technological education for recreational activities. The usage of laboratory experiments is likewise quite low (63%), similar to how 53% of people still do not utilize images, paintings, digital videos, or digital cameras. Regarding digital portfolios and student evaluation, professionals (53%) dispute the role of technology in education with cheating in exams being a prevalent problem to be addressed (Curran, Middleton, and

Doherty, 2011; Keresztury and Cser, 2013). Along with this, the possibility of misinformation and fake information is also very high to trust and adopt (Vajpayee, Sheokand, and Sanghani, 2022) in many aspects of information sharing on internet.

ix. **Mental Health and TE:** One effect is the potential for increased levels of anxiety, stress, and depression. TE encourages students to work independently, and heavy reliance on technology could lead to students feeling isolated and disconnected from their instructors and classmates. This can result in feelings of loneliness and frustration, further leading to increased stress and anxiety (Vajpayee, Devnani, and Sanghani, 2023). The lack of direct contact with teachers and peers can contribute to feelings of depression, particularly for those students who are already predisposed to mental health issues (Devnani, Vajpayee, and Sanghani, 2022). Therefore, it is important to ensure students have access to the necessary support services and are able to interact with their instructors and classmates. Another potential negative effect of TE is the potential for distraction and procrastination (Griffin, 2014). Many students may be drawn to the idea of being able to access their studies and educational materials from virtually anywhere and at any time (Broadbent and Lodge, 2021). This could lead to the students giving into distractions easier or putting off the more difficult or time-consuming tasks.

11.6 DISCUSSION

The provided content explores the impact of TE in higher education, presenting both positive and negative perspectives from educators. Academics widely agree that TE improves academic results and serves as a facilitator for teachers' professional development, reducing their workload (Devnani et al., 2022). There is recognition that TE promotes the development of communication skills and motivates students (Rinekso and Muslim, 2020), but there are concerns about potential negative impacts on mental health due to over-reliance on technology (Kew et al., 2018).

The discussion highlights the innovative potential of TE in introducing concepts and ideas and its role as a tool for enhancing teaching methods (Kapur, 2019). However, there are concerns about reduced interpersonal skills, increased stress, and challenges in time management associated with TE (Hartzell et al., 2016). The content underscores the importance of maintaining a balance between traditional and technological educational tools and emphasizes the need for ongoing research to address mental health concerns and ethical considerations in the use of technology (Vajpayee, Devnani, and Sanghani, 2023).

Gender disparities among participants are mentioned, with 65.3% men and 34.7% women. Despite this disparity, there is a generally positive attitude toward TE, with both genders agreeing on its benefits. The content concludes by emphasizing the need for a balanced approach to TE adoption, avoiding blind application, and ensuring the overall well-being of students and educators (Vajpayee and Ramchandran, 2019).

Traditional education holds a paramount role in society, playing a pivotal part in shaping and nurturing students. It exerts a profound influence on students' intellectual, moral, and social development, imparting experiences, values, and skills crucial for their role as productive contributors to society. Beyond being a wellspring of knowledge, traditional education instills a sense of identity and belonging. It fosters critical thinking,and problem-solving abilities, and cultivates global awareness, ultimately nurturing informed citizens capable of actively shaping the future of their communities (Joshi, Vajpayee, and Mishra, 2005; Vajpayee, 2017a; Vajpayee, 2017b; Vajpayee, 2017c).

Gurukul schools, embodying a traditional educational ethos, renowned for nurturing holistic development in students (Mishra and Vajpayee, 2004; Mishra and Vajpayee, 2008), often find their distinctive approaches missing in the TE system. The unique qualities inherent in traditional education, as evidenced by Gurukul schools, underscore the need for a balanced integration of both traditional and technological approaches to ensure a comprehensive and effective educational experience.

REFERENCES

Al-Rahmi, W. M., Othman, M. S., & Musa, M. A. The improvement of students' academic performance by using social media through collaborative learning in Malaysian. *Asian Social Science* 10, no. 8 (2014): 210–221.

Anderson, T. D., & Garrison, D. R. Critical thinking in distance education: Developing critical communities in an audio teleconference context. *Higher Education* 29, no. 2 (1995): 183–199.

Arasaratnam-Smith, L. A., & Northcote, M. T. Community in online higher education: Challenges and opportunities. *Electronic Journal of e-Learning* 15, no. 2 (2017): 188–198.

Broadbent, J., & Lodge, J. Use of live chat in higher education to support self-regulated help-seeking behaviours: A comparison of online and blended learner perspectives. *International Journal of Educational Technology in Higher Education* 18, no. 1 (2021): 1–20.

Cochran, David S., & Noor O. Borbieva. Collective system design and Industry 5.0: Building community, resilience, and sustainability at Purdue University Fort Wayne. In *International Symposium on Industrial Engineering and Automation*, pp. 258–273. Cham, Switzerland: Springer Nature, 2023.

Criollo-C, S., Luján-Mora, S., & Jaramillo-Alcázar, A. Advantages and disadvantages of M-learning in current education. In *2018 IEEE World Engineering Education Conference (EDUNINE)*, pp. 1–6. IEEE, 2018.

Curran, K., Middleton, G., & Doherty, C. Cheating in exams with technology. *International Journal of Cyber Ethics in Education (IJCEE)* 1, no. 2 (2011): 54–62.

Devnani, S., Vajpayee A., Roy S., & Sanghani, P. Neurological syndrome of anxiety and depression as an outcome of nomophobia. *Journal of Pharmaceutical Negative Results* 13, Special Issue 7 (2022).

Edmonds, C. D. Providing access to students with disabilities in online distance education: Legal and technical concerns for higher education. *American Journal of Distance Education* 18, no. 1 (2004): 51–62.

El-Sabban, F. Incorporating e-mail in teaching activities of the nutrition program at College for Women, Kuwait University: Assessment of a five-year experience. *British Journal of Education, Society & Behavioural Science* 5, no. 1 (2015): 62–72.

Goel, K., Cai, S, X., Engel, L. C., & McCarthy, C. Reassessing globalization: New perspectives on education in the tumult of global change. *Discourse: Studies in the Cultural Politics of Education* 40, no. 5 (2019).

Gordon, N. Flexible pedagogies: Technology-enhanced learning. *The Higher Education Academy* 10, no. 2.1 (2014): 2052–5760.

Grant, P., & Basye, D. Personalized learning: A guide for engaging students with technology. *International Society for Technology in Education* (2014).

Griffin, A. Technology distraction in the learning environment. *SAIS* (2014) Proceedings. 10. https://aisel.aisnet.org/sais2014/10.

Haleem, A., Javaid, M., Qadri, M. A., Suman, R. Goel, K., Xiuying Cai, S., Engel, L. C., & McCarthy, C. (2019). Reassessing globalization: New perspectives on education in the tumult of global change. *Discourse: Studies in the Cultural Politics of Education* 40, no. 5 (2019: 757–759.

Hartzell, J. F., Davis, B., Melcher, D., Miceli, G., Jovicich, J., Nath, T., … Hasson, U. Brains of verbal memory specialists show anatomical differences in language, memory and visual systems. *Neuroimage* 131 (2016).

Joshi, S., A. Vajpayee, & R. C. Mishra. Community mental health and psychological interventions. *Social Science International* 21, no. 2 (2005): 103.

Kantharia N., Vajpayee A. and Sanghani P.(2023). Impact of Socialization and Financial Greed in Endorsing Immoral Decisions among Students, *Journal of Mental Health Issues and Behavior*. ISSN: 2799-1261 Vol: 03, No. 01, Dec 2022 - Jan 2023.

Kapur, R. The significance of ICT in education. *IOSR Journal of Research and Methods* 7, no. 3 (2019): 43–49.

Kärkkäinen, K., & Vincent-Lancrin, S. Sparking innovation in STEM education with technology and collaboration: A case study of the HP Catalyst Initiative OECD iLibrary. (2013).

Keresztury, B., & Cser, L. New cheating methods in the electronic teaching era. *Procedia – Social and Behavioral Sciences* 93 (2013): 1516–1520.

Kew, S. N., Petsangsri, S., Ratanaolarn, T., & Tasir, Z. Examining the motivation level of students in e-learning in higher education institution in Thailand: A case study. *Education and Information Technologies* 23 (2018): 2947–2967.

Knausenberger J, Giesen-Leuchter A, Echterhoff G. Feeling ostracized by others' smartphone use: The effect of phubbing on fundamental needs, mood, and trust. Frontiers in Psychology. 2022 Jul 1;13:883901.

Kumar, U., Singh, A. & Paddakanya, P. Extensive long-term verbal memory training is associated with brain plasticity. *Sci Rep* **11**, 9712 (2021). https://doi.org/10.1038/s41598-021-89248-7

Lauricella, S., & Kay, R. Exploring the use of text and instant messaging in higher education classrooms. *Research in Learning Technology* 21 10.3402/rlt.v21i0.19061 (2013, 21: 19061. http://dx.doi.org/10.3402/rlt.v21i0.1906121:

Mishra, R. C. Vajpayee, A. Sanskrit School in India. In *Educational Theories and Practices from the Majority World,* edited by Pierre R. Dasen and A. Akkari, 245–267. Sage Publication, 2008.

Mishra, R. C., & Vajpayee, A. Les écoles sanskrites en Inde [Sanskrit schools in India]. In *Pédagogies et pédagogues du Sud*, edited by A. Akkari & P.R. Dasen, 207–230. L'Harmattan, 2004.

Nawaz, A., & Khan, M.Z. Issues of technical support for e-learning systems in higher education institutions. *International Journal of Modern Education and Computer Science* 4, no. 2 (2012): 38.

Padmanabhan, A. Advantages and disadvantages of using technology for teaching and learning process in education. *International Journal for Research Trends and Innovation* 5, no. 4 (2020): 138.

Rinekso, A.B., & Muslim, A.B. Synchronous online discussion: Teaching English in higher education amidst the COVID-19 pandemic. *JEES, Journal of English Educators Society* 5, no. 2 (2020) (2):155–162..

Santosh, Sujata, and Santosh Panda. Sharing of knowledge among faculty in a mega open university. *Open Praxis* 8, no. 3 (2016): 247–264.

Sheokand, U. and Vajpayee. A. Critical assessment of RTE Act in India and its comparative statistical assessment with special reference to PTR, corporal punishment, and work-load parameters. *Journal of Pharmaceutical Negative Results* 14, no. 3 (2023): 2063–2071. https://doi.org/10.47750/pnr.2023.14.03.268.

Todd, Wendy F., Chessaly E. Towne, and Judi Brown Clarke. Importance of cantering traditional knowledge and indigenous culture in geoscience education. *Journal of Geoscience Education* 3 (2023): 403–414

Vadher, S, Vajpayee, A., and Sanghani, P. Relational negotiators for compromising ethics and moral values among students: A case of south Gujrat. *Korea Review of International Studies* 16, no. 47 (2023): 60–70.

Vajpayee, A., Devnani S., and Sanghani P. Social isolation, self-aloofness during COVID-19 and its impact on mobile phone addictions. *Korea Review of International Studies* 16, no. 45 (2023), 60–70.

Vajpayee A, Sheokand U., & Sanghani, P. Vulnerability of fake news in the human mind and cognition. *NeuroQuantology* 20, no. 22 (2022): 1406–1413, 247–256, doi:10.48047/nq.2022.20.22. NQ10119. eISSN 1303-5150.

Vajpayee, A., and Sanghani, P. Eternal happiness and endurance of life through Buddhism in Bhutan. *British Journal of Administrative Management* 52, no. 151 (2022): 10–23.

Vajpayee A. and Ramachandran K.K Reconnoitring artificial intelligence in knowledge management. *International Journal of Innovative Technology and Exploring Engineering* (IJITEE) 8, no. 7C (May 2019) 114-117.

Vajpayee, A and Karthick K. K. (2019). Organizational Pyramid and Size as a Moderator Variable in Manufacturing Industries of Bhutan. *International*

Journal of Innovative Technology and Exploring Engineering. International Journal of Innovative Technology and Exploring Engineering (IJITEE). 8 Issue-7S2,-503-509.

Vajpayee A., Mishra R.C., and Dasen P. Spatial encoding: A comparison of Sanskrit- and Hindi-medium schools. In N. Srinivasan, A.K. Gupta, & J. Pandey (Eds.), *Advances in Cognitive Science* (pp. 255–265). New Delhi: Sage Publication, 2008.

Vajpayee, A. and Sanghani P. Eternal happiness and endurance of life through Buddhism in Bhutan. *British Journal of Administrative Management* 52, no. 151 (2022).: 10–23.

Vajpayee, A. Impact of visual literacy skill program on visual literacy of deprived children and non-deprived children. *Likars'ka. Sprava*, 7: 75–80 (2017a). doi:10.31640.

Vajpayee, A. Effect of early intervention on pictorial perception of deprived and non-deprived children. *Indian Journal of Psychological Science* 9, no. 1 (2017b).

Vajpayee, A. Barriers of education for the children of Kharwar Tribe. *International Journal of Indian Psychology* 4, no. 3 (2017c): 112–122.

Zheng, Xiang, Junxiao Bao, and Jinli Wang. Problems and measures of traditional culture education and mental health education in colleges and universities under the New Media Environment. *Journal of Environmental and Public Health* Volume 2022 | Article ID 3370481 | (2022).

/ Chapter 12

Ethical conundrums and privacy concerns in new innovations

A systematic review

Rahul Joshi

12.1 INTRODUCTION: BACKGROUND AND SIGNIFICANCE

A new idea called "Industry 5.0" seeks to revolutionize the industrial sector and create a sustainable, productive, and customized future. Building on earlier industrial revolutions such as mass production, automation, digitalization, and steam power, this new paradigm for manufacturing sector uses the most recent developments in robotics, artificial intelligence (AI), and the Internet of Things (IoT). The "Industry 4.0" movement, which had its start in Germany in 2011, is where Industry 5.0 got its start. The goal of Industry 4.0 was to integrate the digital and physical realms of manufacturing via data analytics, automation, and sensors to enhance productivity. However, mass manufacturing was the focus of Industry 4.0, with little attention paid to sustainability or customization.

The concept of Industry 5.0 is a new and innovative approach to manufacturing that prioritizes sustainability, adaptability, and customization. This is made possible by the combination of emerging technologies, such as the IoT, machine learning, and advanced robotics. By utilizing these technologies, manufacturers can develop highly responsive and flexible manufacturing processes that can produce personalized goods in smaller quantities. The advantages of Industry 5.0 are significant, as it has the potential to revolutionize the industrial sector, drive innovation and development, streamline operations, reduce waste, and meet market demand for new goods and services. Moreover, Industry 5.0 can also contribute to an eco-friendlier manufacturing process by minimizing the impact of production on the environment and promoting a circular economy. However, implementing Industry 5.0 requires significant investments in hardware, software, and skilled personnel who are knowledgeable about cutting-edge technologies. Despite the challenges, many businesses are already investing in Industry 5.0 because the potential benefits outweigh the costs. For instance, the German government has launched the "Industry 4.0" initiative to encourage the adoption of Industry 5.0 and support the digital transformation of the industrial sector.

DOI: 10.1201/9781032677040-12

The potential of emerging technologies in today's world is vast, encompassing cybersecurity, big data and machine learning, AI, autonomous vehicles (AVs), and cloud computing. However, before businesses can capitalize on these breakthroughs in a practical sense, ethical concerns regarding data security and privacy must be addressed. Richards et al. (2023) argued that theories centered on autonomy, fairness, beneficence, non-maleficence, and faithfulness serve as the foundation for ethical thinking and considerations. While ethical reasoning cannot be equated to other disciplines, it does share some characteristics with them. Ethical conclusions cannot be proven with the same certainty as mathematical theorems, but this does not mean that all ethical decisions are equally valid. In fact, most philosophers of science contend that scientific findings are tentative truths that cannot be definitively proven. Regardless of the discipline, certain findings are more likely to be accurate than others. As a result, the idea that ethics has no right or wrong is a common misconception in ethics courses (Jones et al. 2010).

The field of digital ethics has made significant strides by leveraging emerging technologies. Hansson (2017) stated that there is a growing emphasis on early-stage involvement in the technological revolution. Techno Ethics (TE) is a multidisciplinary area of study that draws on various disciplines, including sociology, communication systems, innovation, and ethical theories and principles. For instance, the Internet was never intended to be a secure or private space, but rather a global, free service for everyone (Pawlicka et al. 2021). All cyber-related illegal activities enabled by access to an IT infrastructure are collectively referred to as cybercrime. This includes system disruption, digital identity fraud, illegal data comparison filtration, unauthorized access, and other offenses (Choraś et al 2019). Cybersecurity is the antithesis of cybercrime. Its goal is to help individuals minimize risks in their networks, systems, and data while upholding security and privacy. Both official and unofficial resources, including tools, personnel, infrastructure, services, plans, instruction, and technology, are employed to protect cyberspace (Timmers 2019). As more companies provide information to demonstrate their public commitment to security and ethical principles (Dhirani et al. 2023), ethical standards for new technologies are becoming a topic of increasing concern.

12.2 UNRAVELLING THE COMPLEXITY OF CYBER ETHICS

The advent of new technologies has drastically transformed multiple industries, leading to increased productivity, collaboration, and reliance on these platforms. Nonetheless, these technologies may be misused or hacked, which could cause significant harm to organizations as well as individuals whose data is exposed. This is where ethical considerations about the social contract come into play. A breach of the social contract occurs when a product or service of an organization has a direct impact on public interests, such as privacy, safety, and security. As an example, imagine a healthcare facility that uses an Enterprise Resource Planning (ERP) system to manage its operations. If an IT administrator discovers a vulnerability in the ERP system, they must act quickly to patch it to prevent potential harm. However, the patching process may take up to 12 hours, during which time the ERP system will be out of service. The IT administrator must consider the potential impact on the healthcare facility's in-patient services and treatments. To minimize harm, the IT administrator may choose to implement the patch during off-hours when patient occupancy is lower and surgical operations are not performed. This decision is based on a utilitarian perspective, as it seeks to minimize harm from the administrator's point of view. However, if the patch is not implemented in a timely manner, malicious actors may exploit the vulnerabilities, resulting in severe consequences such as unauthorized access to patient data, theft of personal healthcare information, and the disruption of services via Denial-of-Service attacks, among other possible harms (Furey et al. 2021).

In the realm of cybersecurity ethics, it is crucial to distinguish between the various types of hackers. Those who fall under the "black hat" category engage in unlawful activities, such as exploiting system vulnerabilities to access sensitive data or profit financially. Conversely, "white hat" hackers are ethical professionals who operate within a company's infrastructure to detect and resolve security vulnerabilities by utilizing threat intelligence and conducting penetration testing (Olushola 2018).

The recent case involving the German Christian Democratic Union (CDU) has brought attention to the importance of ethical hacking and the risks associated with zero-day vulnerabilities (Baloo 2023). Grey hat hackers possess comparable levels of expertise to black and white hat hackers, but their motives are distinct. They actively search for security flaws without proper permission or authorization, which contradicts the ethical guidelines established by the cyber community. Grey hat hackers seek financial compensation in exchange for providing complete details about the vulnerabilities they discover. In contrast, red hat hackers, as explained in literature (Avast 2022), utilize offensive tactics to identify and neutralize hostile threat actors. They respond by causing harm to their networks and devices. Red hat hackers are renowned for their ability to penetrate the dark web and engage in offensive operations against malicious actors, commonly known as black hat hackers. Blue hat hackers are cybersecurity experts employed by organizations to conduct penetration testing and enhance their overall cyber defense strategy. Although blue hat hackers possess skills comparable to those of white hat hackers, they differ in the range of services they offer. Green hat hackers are individuals seeking to establish themselves as experts in cyber hacking. They have limited comprehension, practical experience, and technical expertise in the field. Frequently, they visit various online platforms, such as domains and blogs, to seek guidance and ask questions.

The rise of state-sponsored industrial and cyber espionage, along with the emergence of counter-back cyber-attack scenarios, has raised an important question about who has the authority to determine the legality of these activities. However, in this domain, the absence of regulatory frameworks has created several areas of uncertainty. There are different kinds of hackers, each with their own motivations. For instance, in industrial espionage, rivals may try to gain advantages by stealing intellectual property or engaging in eavesdropping activities. Cybercrimes of this kind may be perpetrated by individuals with insider knowledge or who identify as black or grey hat hackers. To safeguard industries from cybercrimes, governments worldwide have established specific authorizations and boundaries for industries to engage in ethical hacking. The primary objective is to protect their data and organizations from potential zero-day attacks and to address legal enforcement dilemmas arising from encryption techniques.

While both cybercrime and cyber terrorism are considered immoral, it is essential to note that these terms include distinct concepts. Instances in which the integrity of vital infrastructures such as the electric grid, water supply, and healthcare facilities is compromised by hostile entities, resulting in direct harm to human life, violate the principles outlined in the social contract theory. Furthermore, such incidents can be classified as instances of cyber terrorism (Dhirani et al. 2021). Currently, there needs to be more established legislation about such circumstances. An ethical quandary emerges in this scenario. Given the circumstances, does the afflicted nation possess a legitimate ethical basis or entitlement to retaliatory hacking against the state responsible for the attack? Bateman et al. (2022) argued, over the previous year, the governments of Russia and Ukraine have initiated a range of state-sponsored assaults. Those that have impacted individuals are classified as instances of cyber terrorism, although there have also been reports of Ukrainian hackers engaging in hack-back operations. The ethical inquiry underscores the pressing need for the implementation of cyber legislation and regulations to address the growing instances of cyber warfare and terrorism.

In response to the increasing incidence of organized cybercrime assaults, several sectors are proactively recruiting and pursuing the expertise of cyber professionals and penetration testers. The objective is to identify vulnerabilities and weaknesses inside their systems and networks before potential exploitation by hostile hackers. The individuals often referred to as "white hat" and "blue hat" hackers (Hawamleh et al. 2020) are expected to adhere to the ethical principles established by the industry. These principles provide a valid basis for penetration testing, cyber forensics, log analysis, and others (Gupta et al. 2022). This exemplifies the significant influence of cyberattacks on the information and operational technology landscape of industries, shedding light on the need to develop and harmonize various cybersecurity and regulatory standards to safeguard environments against cyber data breaches.

12.3 ASPECTS OF ETHICS IN THE CLOUD COMPUTING INDUSTRY

The cloud computing infrastructure has facilitated and delivered auspicious results for the industrial IoT environment. Dhirani et al. (2016) argued that the cloud provides a variety of service models and models. These include technology as a service, software as a service, and technology as a service. A legal contract known as a cloud Service Level Agreement (SLA) is established between a cloud tenant and a service provider to govern the delivery of the promised services. Failure to provide the agreed-upon services may result in a penalty for the vendor (contract nullification is possible) or a renegotiation of the agreement under a revised SLA. Notwithstanding the existence of an SLA governing the quality of service, the most significant ethical and privacy concerns emerge when services are executed on external infrastructures without the awareness or consent of the end-user. Several of the linked conditions and terms mentioned in the SLA generate ambiguity and false statements (Dhirani and Newe 2020).

Around 80% of global industrial data will be analyzed in the cloud by 2025. Such resource requirements are beyond the capabilities of the current models, which necessitate the use of federated and brokerage cloud models. Nevertheless, the federated models exhibit deficiencies in the following areas: business continuity, trust, access management, incident response, and governance, risk, and control (GRC) standards. The NIST Cloud Federation Reference Architecture (NIST SP 500-332) (Faragardi 2017) merely offers a rudimentary comprehension of the functions carried out by various cloud actors (including vendors, carriers, brokers, users, auditors, and others), a delineation of technical and service standards, and recommendations to facilitate the process of adoption (Lee et al. 2020).

Imagine a facility that stores and processes genomic data, analyzing and calculating DNA structures and patterns. This requires significant computational resources, and sometimes the facility needs to share sensitive information for treatment purposes with other research centers in different jurisdictions, utilizing cloud models. If there is any lapse or negligence in cloud security policies or standards, it could have serious consequences for all individuals undergoing treatment, violating the social contract and the General Data Protection Regulation (GDPR). In their publication (Wan et al. 2022), the authors discuss ethical dilemmas about the confidentiality and protection of genomic data. They question whether current compliance and security measures are adequate to safeguard the information given the dynamic nature of emerging technologies. Ethical considerations surrounding cloud computing are influenced by many technological factors, including security, privacy, compliance, and performance metrics.

Considering the expanding implementation of cloud computing across various industries, particularly healthcare, the necessary security, and regulatory controls (GDPR) must be maintained. Due to the misconception that the cloud is an independent and exclusive entity, the security protocols (zero trust, availability, and compliance) implemented in the private and public clouds of an industry may vary, thereby increasing the vulnerability of the environment to cybercriminal breaches. Presently, the standardization organizations have yet to provide or release an interoperable cloud standards platform; thus, establishing visibility, control, and insights would be the only method to mitigate cloud-based risk. A road map for comprehending the distinctions at the SLA levels and connecting the industrial operational environment with the cloud is provided in reference Opara-Martins (2023). However, it is necessary to expand the gap analysis to account for the additional implementation of novel and inventive technologies that aim to address the ethical, social, and privacy concerns.

12.4 ETHICAL PREDICAMENTS ARISING FROM THE AUTONOMOUS VEHICLES

The emergence and implementation of AVs have attracted considerable interest and enthusiasm owing to their potential to transform the field of transportation. These vehicles have the potential to improve road safety, alleviate traffic congestion, and offer enhanced mobility for those with impairments or those who are unable to operate a car. In addition to the benefits above, AVs provide a myriad of ethical quandaries that need societal deliberation considering ongoing technological advancements. This extensive discourse will thoroughly investigate the many ethical considerations presented by autonomous cars, thoroughly analyzing each one and suggesting alternative remedies.

12.4.1 Trolley Problem

The Trolley Problem is a well-recognized moral thought exercise that has garnered significant attention and sparked much discourse about AVs. This circumstance compels us to contemplate the appropriate action for an AV when confronted with an inevitable collision. When a kid unexpectedly enters the roadway, an AV is faced with either swerving, which may pose a risk to its occupants, or colliding with the child. In such a situation, what would be the optimal choice for the AV? The ethical quandary is on whom or what the AV should prioritize: the safety of its passengers, who may bear no responsibility for the accident, or the safety of pedestrians and other individuals using the road. Resolving this problem is a complex task as it necessitates making a predetermined selection among many types of injury. One plausible strategy is using social agreement to define default configurations for AVs. Through the implementation of surveys and the collection of information from many stakeholders, it is possible to ascertain the values and priorities that need to govern decision-making about AVs. Nevertheless, this methodology fails to address the issue of personalized configurations – should individuals who own autonomous cars have the freedom to modify the moral algorithms of their vehicles? Achieving an optimal equilibrium between soliciting public feedback and implementing customization continues to pose a significant problem (Wiseman et al. 2018).

12.4.2 Obligation and liability

In situations involving AVs, determining accountability can be a complex ethical quandary. Unlike traditional vehicles, where the driver is typically held responsible, the manufacturer, software developer, or owner of the AV may bear the burden. AVs rely on a combination of software and hardware to function. Yazdanpanah et al. (2023) argued that concerns surrounding software, such as faulty decision-making algorithms, could be attributed to the developer, while problems with hardware, such as sensor malfunctions, may fall on the manufacturer. Additionally, if an AV is involved in an accident while in a mode where the human driver has some control, assigning responsibility becomes even more challenging. To address this issue, it is crucial to establish legal and regulatory frameworks that outline the roles and responsibilities of all relevant parties. These frameworks should support innovation in the AV industry while ensuring accountability for accidents. However, creating and implementing consistent frameworks across different jurisdictions will be a challenge.

12.4.3 Data privacy

Advanced self-driving vehicles are equipped with sensors that gather a vast amount of information about their environment and passengers. This data is critical for the safe operation and decision-making of these vehicles. However, the collection, storage, utilization, and sharing of this information can raise ethical concerns. Hataba et al. (2022) stated that AVs have the potential to function as advanced surveillance systems, monitoring the activities, location, and conversations of their occupants. This data could be used for commercial purposes, such as selling it to third parties or using it for targeted advertising. These issues raise concerns about privacy, consent, and control over personal data. To address these concerns, AV software manufacturers must implement robust data protection and encryption mechanisms. Individuals should have control over their data, including the ability to decide how it is used, shared, or deleted, in compliance with relevant privacy laws and regulations. Transparency regarding data collection practices is essential, and users must be informed about the data collected by AVs and how it will be used.

12.4.4 Hacking and cybersecurity

The susceptibility of AVs to cyber assaults and malware presents a significant ethical dilemma. AVs are complex computer systems with multiple entry points for potential attacks, which makes them vulnerable to exploitation by malicious actors for various reasons, including causing damage or obtaining sensitive data. Ensuring the safety and security of AV occupants and other road users is an ethical imperative. A cyber-physical assault on AVs could have life-threatening consequences. Therefore, it is crucial to establish and uphold resilient cybersecurity protocols to protect these vehicles efficiently. To achieve this, AV developers and manufacturers must implement robust authentication, encryption, and intrusion detection systems. Collaboration with cybersecurity specialists and ethical hackers is also vital for identifying and addressing vulnerabilities. Abou Abdalla and Goyal (2022) argued that governments may conduct security audits and evaluations, as well as regulate and supervise the cybersecurity practices of AV manufacturers. Additionally, educating and raising public awareness about the potential cybersecurity risks of AVs can encourage their responsible use and secure operation.

12.4.5 Bias and discrimination

AVs rely on machine learning algorithms that are trained on vast datasets to make real-time decisions. However, if these datasets contain prejudices related to socioeconomic status, race, gender, or any other factor, AVs may unintentionally make biased decisions. This raises serious ethical concerns about fairness and equity in the implementation of AV technology (Janatabadi and Ermagun 2022). For example, when an AV consistently shows hesitation or aggression toward drivers from a particular demographic group while giving the right of way to drivers from another group, it perpetuates discrimination. Although this discrimination may not be intentional, it could be attributed to the biases present in the training data. To tackle this issue, it is essential to have a meticulous data purification process and adhere to equity in algorithm development. AV manufacturers must diligently identify and remove biases from their training data. Moreover, continuous surveillance and testing must be put in place to ensure there is no discriminatory behavior. Additionally, the development and validation of algorithms should be made transparent to promote accountability and build trust.

12.4.6 Ethical hacking and malicious use

Ethical hacking is an essential element in safeguarding AVs. However, it also poses a difficult ethical dilemma by potentially exposing those vulnerabilities to malicious exploitation while simultaneously revealing vulnerabilities that could be improved for AV cybersecurity. White hat hackers engage in ethical hacking by proactively identifying vulnerabilities in AV systems. Their assistance helps manufacturers and developers in correcting these susceptibilities, thereby bolstering the resilience of AVs against assaults. Nonetheless, this methodology gives rise to apprehensions concerning the possible exploitation of these detected weaknesses by malevolent entities (Nobles 2023). AVs, which are intricate computer systems, are vulnerable to hacking endeavors such as remote compromise, data pilferage, and manipulation of decision-making algorithms. As a result, they present substantial threats to both security and privacy. It is crucial to strike a balance between the necessity for robust security measures, which includes ethical hacking, and protection against malicious intent. Securing software development practices, regulatory supervision, and explicit guidelines are imperative to mitigate vulnerabilities. Advocacy for best cybersecurity practices, ongoing public awareness, and vigilant surveillance are imperative to safeguard AVs against the perils linked to malevolent cyberattacks and maintain their status as secure and reliable modes of transportation.

12.5 UNDERSTANDING ETHICAL ARTIFICIAL INTELLIGENCE

Siau and Wang (2020) argued that artificial intelligence (AI) has made remarkable progress in various fields such as AVs, medical diagnosis, and machine vision, demonstrating its potential for advancing human welfare, economic growth, and societal progress. However, AI-based technologies also pose significant risks to societies, industries, and communities due to issues such as lack of interpretability, inaccurate data, compliance with data protection and privacy regulations, and ethical concerns. The creation of AI that adheres to moral and ethical standards is one of the most significant challenges in this field. To address this issue, sectors must consider both AI ethics and methods for developing ethical AI. AI ethics involves examining the moral principles, regulations, standards, and legislation that apply to AI, adhering to fundamental tenets such as privacy, transparency, respect for human values, fairness, safety, and accountability. These principles are like those outlined in the European GDPR (EU GDPR 2022). The EU AI Act, adopted in 2022, seeks to establish a legal framework for AI to foster confidence and mitigate potential damage caused by the technology. However, some members of the European Parliament have expressed concerns about evaluating fundamental rights for high-risk users. The AI Act mandates a comprehensive risk impact assessment plan for diverse threat scenarios and potential exposures such as compliance, AI-cybersecurity, etc. It is crucial to understand the underlying principles of AI ethics and potential avenues for the development of ethical AI, especially given the ongoing development of the AI Act.

12.5.1 Fundamentals of ethical AI

1. Transparency is an essential tenet of ethical AI. It involves guaranteeing the explainability of AI systems and interpreting the decision-making processes underlying AI algorithms. Transparent AI fosters confidence and empowers users to understand the reasoning and process behind a specific decision.
2. It is critical to ensure impartiality in AI systems to prevent prejudice and discrimination. AI algorithms must maintain objectivity about demographics such as ethnicity, gender, and socioeconomic standing. It is imperative to mitigate biases in both algorithmic decision-making and training data.
3. Accountability pertains to the capacity to attribute responsibility in an AI system malfunction. It entails determining accountability for AI deployment, development, and potential adverse outcomes. It is critical to establish unambiguous channels of accountability to effectively address ethical breaches or errors that may occur within AI systems.
4. User privacy protection is an essential ethical consideration. AI systems frequently handle substantial volumes of personal data; therefore, it is critical to enforce robust data protection protocols, obtain informed consent from users, and comply with applicable privacy regulations to safeguard sensitive information about individuals. Google's project Nightingale and Ascension (Prainsack 2020) lawsuits, which arose from the collection of personal data and gave rise to privacy apprehensions regarding data sharing and the application of AI, should serve as a cautionary tale. There are numerous quandaries concerning the practicality of AI. For instance, integrating AI into self-driving vehicles has generated significant ethical considerations. When the software was developed using a utilitarian approach, it would select the option with the fewest casualties in the event of a collision. However, when programmed according to the social contract theory, the AI could not make decisions and repeatedly iteratively searched for predetermined conditions, ultimately leading to an accident because it failed to identify the required conditions. Developing AI to exhibit human-like reasoning and moral and ethical behavior is one of the most significant obstacles; nevertheless,

the expanding autonomous and self-driving industry means no turning back. Therefore, developing comprehensive regulations and standards would be the only way to regulate ethical issues associated with AI.

5. AI systems must be developed and evaluated to guarantee that they function securely. This encompasses strategies to avert robotic mishaps, reduce potential hazards, and establish procedures for managing unforeseen or potentially detrimental AI conduct. AIs and healthcare are sectors where safety is of the utmost importance.

12.5.2 Why Ethical AI Is Crucial?

AI has the potential to significantly transform several aspects of human existence, including domains such as healthcare, economics, transportation, and entertainment. De Cremer and Narayanan (2023) argued that the integration of ethical issues into the development of AI systems is becoming more imperative due to their fast growth. Incorporating ethical principles and societal values into AI systems, often called ethical AI, has emerged as a crucial element within technology. This comprehensive investigation examines the fundamental rationales behind the need for ethical AI, focusing on the urgent requirement for responsible breakthroughs in AI that adhere to technical accuracy and scientific rigor.

1. **Fairness and Bias Mitigation:** An essential aspect of ethical AI is addressing prejudice and striving for justice within AI systems. Machine learning algorithms acquire knowledge from extensive datasets, which might unintentionally include biases that exist in the actual world. If left unattended, these biases can materialize as discriminatory practices in the decision-making processes of AI, therefore affecting persons based on their race, gender, age, or other legally safeguarded attributes (Borenstein and Howard 2021). The use of scientific rigor is crucial in the development of AI algorithms that possess the capability to identify and rectify biases, hence guaranteeing that AI systems provide fair and just results. In addition, resolving difficulties related to fairness necessitates the examination of sophisticated methodologies such as adversarial networks, data re-weighting, and counterfactual fairness. These approaches need severe scientific investigation.

2. **Privacy Protection:** AI systems often handle extensive volumes of personal data, giving rise to substantial issues over privacy. The incorporation of sophisticated cryptography and privacy-preserving methodologies, such as homomorphic encryption and differential privacy, is essential for the establishment of ethical AI systems. These methodologies facilitate the functioning of AI systems with

confidential information while also safeguarding the privacy of individuals. The presence of scientific competence is crucial for the successful development and implementation of privacy-enhancing technology.

3. **Accountability and Transparency:** Ethical AI fosters a sense of responsibility and openness in the development and decision-making procedures about AI. This involves the development of AI models that not only demonstrate excellent functionality but also enable stakeholders to examine and comprehend their inner workings critically. To ensure the interpretability of AI systems, it is essential to adopt scientifically rigorous procedures, such as the utilization of model-agnostic interpretability techniques and the implementation of explainable AI. These methodologies provide the thorough evaluation of AI outputs by developers, regulators, and end-users, ensuring organizational accountability for any inadvertent repercussions.

4. **Safety and Security:** It is paramount to prioritize the implementation of stringent safety and security measures in the development of AI technologies. The scientific community plays a crucial role in establishing robust protocols that safeguard AI systems against malicious attacks, breaches of data security, and disastrous malfunctions. This involves examining the resilience of AI systems against adversarial assaults, creating secure architectures, and conducting comprehensive threat modeling. In sectors that are particularly vital, such as AVs and healthcare, the importance of AI safety is magnified by the potential for errors to result in life-threatening situations.

5. **Human Well-Being:** The overarching goal of ethical AI is to optimize and augment the overall welfare of human beings. Scientific inquiry plays a crucial role in evaluating the societal ramifications of AI applications, including its effects on employment, social dynamics, and the general well-being of individuals. It is essential to comprehend these consequences and use AI to augment human well-being. In healthcare, the implementation of AI-driven diagnostics has the potential to enhance patient outcomes. However, it is crucial to approach the use of these technologies with ethical concerns in mind. This is necessary to prevent the displacement of human knowledge and empathy by AI systems (Sharma and Bhardwaj 2024).

6. **Public Trust:** The establishment and broad use of AI technology heavily relies on the crucial factor of public trust. The use of ethical practices in the field of AI serves to enhance trust among stakeholders, as it showcases a solid dedication to the responsible development and deployment of AI technologies. It is essential for the technical and scientific community to proactively participate in interactions with the public, policymakers, and relevant stakeholders to promote transparency and facilitate a reciprocal exchange of information. This

fosters a sense of assurance that AI technologies are being developed with a strong focus on the welfare and well-being of society.

7. **Legal and Regulatory Compliance:** The ethical considerations of AI are intricately connected to the adherence to legal and regulatory frameworks. The laws and regulations about AI are in a constant state of flux, necessitating the presence of scientific competence to traverse the intricate terrain effectively. Organizations and developers need to ensure that their AI systems comply with legal obligations while also considering broader ethical standards. Compliance serves the dual purpose of mitigating legal liabilities and bolstering ethical obligations.

8. **Long-Term Viability:** Failure to take ethical factors into account might result in negative consequences such as public reaction, legal disputes, and harm to reputation, thereby posing a threat to the long-term sustainability of AI initiatives. The prevention of these problems is facilitated by the conscientious development of AI, which is underpinned by scientific and technological rigor. The use of appropriate measures guarantees the long-term sustainability, ethical integrity, and societal benefits of AI advancements.

9. **Global Impact:** The use of AI technology extends beyond national borders, hence necessitating international collaboration and agreement on the responsible development of AI. Establishing ethical norms and standards transcending multiple cultural settings is paramount, necessitating scientific inquiry and cooperation across international boundaries. This comprehensive strategy aids in the development and implementation of AI in a way that is advantageous for the whole of humanity while simultaneously mitigating any adverse effects.

10. **Ethical Innovation:** Integrating ethical concepts into AI systems promotes the development of innovative solutions that adhere to ethical standards. This entails a comprehensive scientific investigation of the ethical ramifications of developing technology and formulating innovative approaches that align with established ethical principles. Ethical innovation is vital in guiding AI advancement toward accountable and beneficial results.

12.6 ETHICAL ISSUES WITH MACHINE LEARNING AND MASSIVE DATASETS

Big data and machine learning together constitute a powerful new paradigm in technology that gives unmatched potential but also raises several complex ethical issues. In the following discussion we will examine the critical ethical issues that arise from the combination of big data and machine learning and provide precise scientific and technological explanations for each (Sharifani and Amini 2023). These issues include data ownership and control, social effects, privacy, bias, openness, and responsibility.

1. **Privacy Concerns:** One of the most critical ethical issues in the field of big data and machine learning is privacy. The vast collection, processing, and analysis of personal data has raised fundamental concerns about human privacy, permission, and data security. These concerns have both scientific and technological roots.
    a. **Data Collection:** Complex data acquisition methods and techniques are required due to the massive gathering of personal data from many sources, including social media, IoT devices, and online activities. Scientific investigations must explore the finer points of data collecting, which often include digital footprints and data sensors, while also looking at the privacy consequences.
    b. **Informed Permission:** Scientific examination is required due to the need for informed permission for the collection and use of data. Scientific study on efficient consent procedures is necessary to address the difficulty of educating people and understanding their agreement considering the complicated network of data gathering from numerous sources.
    c. **Data Security:** Advanced encryption methods, robust access control, and cutting-edge cybersecurity procedures are essential for ensuring data security and strengthening personal data protection measures. Scientific and technological developments are essential to safeguarding people's data against breaches and unwanted access.
2. **Bias and Fairness:** The domains of big data and machine learning are characterized by pervasive concerns of bias and fairness, which, therefore, need rigorous scientific and technological examination. The introduction of bias into the data and computational processes has the potential to result in outputs that are unfair and discriminating. **Relevant factors encompass the following:**
    a. **The Influence of Biased Training Data:** Machine learning models, when trained on historical data, tend to adopt the biases that are present within these data sources. The mitigation of bias requires a rigorous examination of data sources and de-biasing methodologies, such as applying re-sampling and re-weighting approaches to eradicate bias.
    b. **Discriminatory Outcomes:** Discrimination in machine learning models might be unintentional and become evident via technical manifestations such as classification mistakes. An essential component in assessing and addressing discriminatory effects is a thorough and rigorous scientific analysis of fairness metrics and

assessment methodologies. This necessitates the development of techniques for measuring classification fairness and ensuring statistical parity.

c. **Fairness Measures:** Establishing fairness requires rigorous scientific and technological methodologies, including sophisticated statistical analysis and machine learning assessment. The development and validation of metrics for assessing fairness in prediction models need careful design, considering factors such as differential impact and equal opportunity. Additionally, the incorporation of modern statistical methods is essential for the implementation of fairness-aware modeling.

3. **Transparency and Explainability:** The importance of openness and explainability is underscored by the lack of transparency in several machine learning algorithms and models. The growing complexity of these algorithms necessitates scientific and technical endeavors.

   The primary characteristics include the following:

   a. **Accountability:** The notion of accountability necessitates scientific inquiry into the interpretability of models, with a particular emphasis on elucidating the decision-making mechanisms inside machine learning models from a technology standpoint. Analytical tools such as LIME (Local Interpretable Model-Agnostic Explanations) and SHAP (Shapley Additive exPlanations) provide techniques for investigating the underlying workings of complex models.

   b. **Legal Compliance:** Ensuring adherence to legal obligations, such as the GDPR, which mandates the ability to understand and challenge automated decisions, requires the implementation of technological mechanisms that offer visibility into the forecasts generated by the model. Incorporating methodologies such as feature significance, sensitivity analysis, and model interpretation has considerable importance in guaranteeing conformity with regulatory requirements.

   c. **Trust and Adoption:** The establishment of trust in machine learning systems, a crucial technical challenge, necessitates the development and integration of interpretable models, visualizations, and explanatory aids. The ongoing advancements in technical innovations contribute to the increased trust and public acceptance of interpretable machine learning models, such as decision trees and rule-based systems.

4. **Accountability and Responsibility:** The distinction between accountability and responsibility within big data and machine learning gives rise to complex discourse. The scientific and technical dimensions

comprise a range of factors and aspects that pertain to the field of study or practice under consideration. These dimensions involve the use of scientific principles, methodologies, and techniques, as well as the utilization of specialized knowledge and expertise to address and solve complex problems.

a. **Data Ownership:** The determination of data ownership and its accompanying duties is contingent upon legal, technological, and contractual contexts. Various stakeholders, including data collectors, processors, and people, are involved in the intricate network of data ownership, protection, and ethical use. Data ownership is supported by technical solutions, such as blockchain, which decentralize control over data.

b. **Algorithm Creators:** It has an ethical obligation that necessitates conducting scientific examinations and implementing supervision mechanisms to evaluate the efficacy of their models. The implementation of algorithm auditing and validation, facilitated using ethical AI toolkits and responsible AI frameworks, serves to enhance the ethical progression and implementation of algorithms.

c. **Legal Frameworks:** The establishment of legal frameworks to assign responsibility in instances of ethical violations necessitates the use of scientific analysis and the development of rules and laws. Frameworks such as the GDPR establish the specific responsibilities and requirements that data controllers and processors must adhere to, thereby requiring the inclusion of scientific expertise to ensure legal conformity.

5. **Impact on Society:** The social implications of big data and machine learning are diverse and are characterized by intricate technological intricacies. The examination of this ethical problem involves the following aspects:

a. **Job Displacement:** The technological implications of job displacement are evident in the automation of jobs facilitated by machine learning and AI. The advent of advanced technologies has created a need for scientific inquiry into the socioeconomic consequences of automation, including retraining programs and the dynamics of the labor market.

b. **Economic Inequality:** A comprehensive examination of economic inequality necessitates incorporating modeling and simulations to evaluate the ramifications of AI and machine learning on differences in income. Machine learning algorithms used in socioeconomic research and wealth distribution provide valuable means for the technical assessment of economic inequality.

   c. **Discriminatory Practices:** The detection and reduction of discriminatory practices enabled by big data and machine learning include the technical analysis of fairness-aware algorithms and auditing tools. The process of examining and addressing unfair results requires thorough data analysis and the creation of models that priorities fairness.

   d. **The Ethical Utilization of Data:** The ethical application of data for the betterment of society is inherently connected to the technological implementation of data analytics for the benefit of the public. Using machine learning methods in social research and policy formation requires a proficient understanding of technical skills such as data analysis, feature engineering, and model validation.

6. **Data Ownership and Control:** The complex issue of data ownership and management encompasses both legal and technological aspects. One potential approach to address the issue of data ownership and control is the use of blockchain technology. This technology offers people the ability to exercise decentralized control over their data via the implementation of mechanisms such as cryptographic key management, data access authorization protocols, and decentralized data control mechanisms.

12.7 REGULATION AND LEGAL ISSUES

This section analyses new technologies from the standpoint of policy, privacy, and legality. Emerging Technologies include the forefront of technological advancements that drive digital transformation. Their significant innovation and cutting-edge nature characterize these technologies. However, it is crucial to address and minimize any threats and risks associated with these technologies to set ethical norms that facilitate their responsible use. The Precautionary Principle (PP) is a framework that advocates for the exercise of caution throughout the development of novel technologies, particularly in cases where there exists a possibility of adverse effects on the environment and human health (Kendal 2022). According to those who argue that the PP stifles innovation by imposing impractical conditions on the safety of emerging technology, there has been significant controversy about this safety principle, surpassing that of any other (Hansson 2020).

To uphold and respect human values in the design process, a widely used technique is known as "value-sensitive design." This approach seeks to identify relevant human values throughout technology development and research (Umbrello and de Poel 2021). The interaction between individuals, technology, and the environment entails several supplementary values at stake, including but not limited to privacy preservation, environmental sustainability, and responsibility. Technology Assessment (TA), along with the PP, is widely recognized as one of the most prominent methodologies for managing uncertainties. Technological Ethics is a methodology that involves the critical evaluation and thorough investigation of concepts, designs, plans, or visions about future technology. This approach aims to generate and evaluate potential information about the future implications and ramifications of technological advancements. The user's text needs to contain information to rewrite academically (Grunwald 2020). The new and emerging strategic technologies (NEST) ethical framework achieves three key objectives. The first phase involves defining the commitments and expectations linked to a transformative framework. Subsequently, this document delineates significant arguments that may be posited in favor of or against the predictions above, particularly those about efficiency and effectiveness. Additionally, the document enumerates other conventional ethical concerns, including individual rights, harm, obligations, fair allocation, and subjective well-being, among other factors. Finally, the paper acknowledges a range of arguments and counterarguments on the benefits and drawbacks of the technology, which might be used to anticipate the trajectory of moral deliberations around technical progress.

The proliferation of developing technologies has raised concerns over privacy and ethics, with potential adverse consequences outweighing the benefits. These challenges can only be effectively addressed by the establishment of an obligatory ethical code of conduct that aligns with existing norms and legislation, therefore ensuring compliance within industries. The European Union's (EU's) digital strategy has been actively engaged in the formulation of standards and legislation about emerging technologies. Several pieces of legislation that pertain to the subject matter include the EU Cyber Resilience Act, the Network, and Information Security Directive (NIS2D) (Gonzalez Pouso 2023), the GDPR, the AI Act, the Digital Markets and Services Act, the Digital Operational Resilience (DORA, 2022), the EU Cybersecurity Strategy (Bendiek and Kettemann 2021), the EU Cybersecurity Act (Pupillo 2018), and the EU Toolbox (Bieber and Maiani 2014), among others. The primary goals of this legislation are to foster trust, transparency, authenticity, accountability, responsibility, and ease while promoting increased corporate activities within the EU that are facilitated by secure and protected data exchange. Companies need to possess a comprehensive understanding of digital capabilities and interoperability across enabling technologies and the frameworks above. It is essential to establish a comprehensive risk and incident assessment framework for each of these technologies as they exhibit variations in their respective features and functionalities. To establish an effective industrial environment, each technology needs to adhere to the principles of data privacy, security, and ethics, notwithstanding the distinct functions they serve. The legislation has integrated ethical considerations into the privacy and security environment, mitigating difficulties.

Nevertheless, in the contemporary day, these rules may serve as a guiding framework for establishing a safe and morally sound atmosphere. Nonetheless, they do not possess the ability to ensure the complete prevention of a data breach. In this context, it is incumbent upon end-users to establish alignment and use optimal methodologies within their own commercial or industrial settings.

12.8 CONCLUSION

In recent decades, the industrial sector has seen substantial changes, transitioning from conventional approaches to Industry 4.0 and Industry 5.0. This paradigm change has presented many possibilities and problems, thoroughly examined, and analyzed in this research endeavor. The research conducted successfully identified the fundamental attributes of Industry 5.0, including heightened levels of customization and adaptability, greater cooperation between humans and machines, and higher standards of safety and sustainability. Nevertheless, the implementation of Industry 5.0 presents many technological, human, socioeconomic, ethical, and legal obstacles. Addressing these difficulties is essential to fully harness the promise of Sector 5.0 and guarantee a sustainable future for the sector. The investigation also examined the prospective effects of Industry 5.0 on production and society. The advantages of Industry 5.0 are many, yet it is imperative to guarantee that implementing this technology does not further amplify existing social and economic disparities. Policymakers need to guarantee the implementation of Industry 5.0 in a manner that confers advantages to all stakeholders, including workers, customers, and the environment. The research findings indicate that Industry 5.0 provides a notable prospect for the manufacturing sector to augment its operational effectiveness, output, and ecological viability.

Nonetheless, the successful implementation of Industry 5.0 necessitates a collaborative endeavor, including policymakers, industry participants, and scholars, to effectively tackle the diverse array of obstacles and secure a viable and enduring trajectory for the sector. In brief, this research underscores the prospective advantages of Industry 5.0 while concurrently recognizing the obstacles that need resolution to facilitate a seamless progression into this emerging epoch of production. Adopting a comprehensive strategy that considers the technological, human, ethical, legal, and socioeconomic dimensions of Industry 5.0 is necessary. Through the implementation of this approach, it becomes possible to fully comprehend the inherent capabilities of this technology and guarantee a viable and enduring trajectory for the industrial sector. Given the dynamic nature of technological advancements, establishing a definitive policy or standard for ethical considerations in the field becomes challenging. Each ethical challenge in technology has unique characteristics that need distinct social, ethical, and legal resolutions. This chapter presents an examination of several facets of the process of ethical decision-making. It also offers valuable insights for those who are new to the field, focusing on the establishment of an ethical, legal, and standardized industrial environment that incorporates modern technology.

REFERENCES

Abou Abdalla, Kaiss Omar, and S. B. Goyal. "Autonomous vehicles: Improving cyber security." *International Journal of Advanced Research in Technology and Innovation* 4, no. 1 (2022): 118–126.

Avast. Hacker Types: Black Hat, White Hat, and Gray Hat Hackers. Last modified October 12, 2022. www.avast.com/c-hacker-types.

Baloo, J. White hat hacking, and the CDU case in Germany. https://blog.avast.com/white-hat-hacking-and%20cdu-avast?_ga=2.12892798.811119647.1665144727-1259669144.1665144727 (accessed on 5 January 2023).

Bateman, Jon, Nick Beecroft, and Gavin Wilde. What the Russian invasion reveals about the future of cyber warfare. 2022. https://carnegieendowment.org/posts/2022/12/what-the-russian-invasion-reveals-about-the-future-of-cyber-warfare?lang=en¢er=global

Bendiek, Annegret, and Matthias C. Kettemann. "Revisiting the EU cybersecurity strategy: A call for EU cyber diplomacy." *Stiftung Wissenschaft und Politik (SWP)* 16 (2021): 1–8.

Bieber, Roland, and Francesco Maiani. "Enhancing Centralized Enforcement of EU law: Pandora's Toolbox." *Common Market Law Review* 51 (2014): 1057.

Borenstein, Jason, and Ayanna Howard. "Emerging challenges in AI and the need for AI ethics education." *AI and Ethics* 1 (2021): 61–65.

Choraś, Michał, Marek Pawlicki, and Rafał Kozik. "The feasibility of deep learning use for adversarial model extraction in the cybersecurity domain." In *Intelligent Data Engineering and Automated Learning – IDEAL 2019: 20th International Conference, Manchester, UK, November 14–16, 2019, Proceedings, Part II* 20, pp. 353–360. Springer International Publishing, 2019.

De Cremer, David, and Devesh Narayanan. "How AI tools can – and cannot – help organizations become more ethical." *Frontiers in Artificial Intelligence* 6 (2023): 1093712.

Dhirani, Lubna Luxmi, Eddie Armstrong, and Thomas Newe. "Industrial IoT, cyber threats, and standards landscape: Evaluation and roadmap." *Sensors* 21, no. 11 (2021): 3901.

Dhirani, Lubna Luxmi, Noorain Mukhtiar, Bhawani Shankar Chowdhry, and Thomas Newe. "Ethical dilemmas and privacy issues in emerging technologies: A review." *Sensors* 23, no. 3 (2023): 1151.

Dhirani, Lubna Luxmi, and Thomas Newe. "Hybrid cloud SLAs for industry 4.0: Bridging the gap." *Annals of Emerging Technologies in Computing (AETiC), Print ISSN* (2020): 2516–0281.

Dhirani, Lubna Luxmi, Thomas Newe, and Shahzad Nizamani. "Tenant-vendor and third-party agreements for the cloud: Considerations for security provision." *International Journal of Software Engineering and Its Applications* 10, no. 12 (2016): 449–460.

Digital Operational Resilience Act (DORA) - Regulation (EU) 2022/2554. Accessed May 29, 2024. www.digital-operational-resilience-act.com/

EU GDPR. "Complete guide to GDPR compliance." 2022. https://gdpr-info.eu/art-32-gdpr/

Faragardi, Hamid Reza. "Ethical considerations in cloud computing systems." In *Proceedings*, vol. 1, no. 3, p. 166. MDPI, 2017.

Furey, Heidi, Scott Hill, and Sujata K. Bhatia. *Beyond the Code: A Philosophical Guide to Engineering Ethics*. Routledge, 2021.

Gonzalez Pouso, Virginia. "The Dutch critical infrastructure under NIS2 directive: A cybersecurity risk-management approach." Master's thesis, 2023.

Grunwald, Armin. "The objects of technology assessment. Hermeneutic extension of consequentialist reasoning." *Journal of Responsible Innovation* 7, no. 1 (2020): 96–112.

Gupta, Meenu, Akash Gupta, and Simrann Arora. "Addressing the security, privacy, and trust issues in IoT-enabled CPS." In *Handbook of Research of Internet of Things and Cyber-Physical Systems*, pp. 433–452. Apple Academic Press, 2022.

Hansson, Sven Ove. "How extreme is the precautionary principle?" *NanoEthics* 14, no. 3 (2020): 245–257.

Hansson, Sven Ove, ed. *The Ethics of Technology: Methods and Approaches*. Rowman & Littlefield, 2017.

Hataba, Muhammad, Ahmed Sherif, Mohamed Mahmoud, Mohamed Abdallah, and Waleed Alasmary. "Security and privacy issues in autonomous vehicles: A layer-based survey." *IEEE Open Journal of the Communications Society* 3 (2022): 811–829.

Hawamleh, A. M. A., Almuhannad Sulaiman M. Alorfi, Jassim Ahmad Al-Gasawneh, and Ghada Al-Rawashdeh. "Cyber security and ethical hacking: The importance of protecting user data." *Solid State Technology* 63, no. 5 (2020): 7894–7899.

Janatabadi, Fatemeh, and Alireza Ermagun. "Empirical evidence of bias in public acceptance of autonomous vehicles." *Transportation Research Part F: Traffic Psychology and Behaviour* 84 (2022): 330–347.

Jones, A., A. McKim, and M. Reiss. *Towards Introducing Ethical Thinking in the Classroom: Beyond Rhetoric Edits in Ethics in the Science and Technology Classroom*. Netherlands: A New Approach to Teaching and Learning, 2010.

Kendal, Evie. "Ethical, legal and social implications of emerging technology (ELSIET) symposium." *Journal of Bioethical Inquiry* 19, no. 3 (2022): 363–370.

Lee, Craig A., Robert B. Bohn, and Martial Michel. "The NIST cloud federation reference architecture 5." *NIST Special Publication* 500 (2020): 332.

Nobles, Calvin. "Offensive artificial intelligence in cybersecurity: Techniques, challenges, and ethical considerations." In *Real-World Solutions for Diversity, Strategic Change, and Organizational Development: Perspectives in Healthcare, Education, Business, and Technology*, pp. 348–363. IGI Global, 2023.

Olushola, OMOYIOLA Bayo. "The legality of ethical hacking." *IOSR Journal of Computer Engineering (IOSR-JCE)* 20, no. 1 (2018): 61–63.

Opara-Martins, Justice. "Perspective chapter: Cloud lock-in parameters – service adoption and migration." In *Edge Computing-Technology, Management and Integration*. IntechOpen, 2023.

Pawlicka, Aleksandra, Michał Choraś, Rafał Kozik, and Marek Pawlicki. "First broad and systematic horizon scanning campaign and study to detect societal and ethical dilemmas and emerging issues spanning over cybersecurity solutions." *Personal and Ubiquitous Computing* 27 (2021): 193–202.

Prainsack, Barbara. "The political economy of digital data: Introduction to the special issue." *Policy Studies* 41, no. 5 (2020): 439–446.

Pupillo, Lorenzo. "EU cybersecurity and the paradox of progress." *CEPS Policy Insight* 2018/06 6 (2018): 1–6. https://cdn.ceps.eu/wp-content/uploads/2018/02/PI2018_06_LP_ParadoxProgress.pdf.

Richards, Deborah, Ravi Vythilingam, and Paul Formosa. "A principlist-based study of the ethical design and acceptability of artificial social agents." *International Journal of Human-Computer Studies* 172 (2023): 102980.

Sharifani, Koosha, and Mahyar Amini. "Machine learning and deep learning: A review of methods and applications." *World Information Technology and Engineering Journal* 10, no. 07 (2023): 3897–3904.

Sharma, Harshwardhani and Saket Kumar Bhardwaj. "Ethical Explorations: AI's Role in Shaping the Journalism Ecosystem Across Asia" *New Global Studies* (2024). https://doi.org/10.1515/ngs-2024-0002.

Siau, Keng, and Weiyu Wang. "Artificial intelligence (AI) ethics: Ethics of AI and ethical AI." *Journal of Database Management (JDM)* 31, no. 2 (2020): 74–87.

Timmers, Paul. "Ethics of AI and cybersecurity when sovereignty is at stake." *Minds and Machines* 29 (2019): 635–645.

Umbrello, Steven, and Ibo Van de Poel. "Mapping value sensitive design onto AI for social good principles." *AI and Ethics* 1, no. 3 (2021): 283–296.

Wan, Zhiyu, James W. Hazel, Ellen Wright Clayton, Yevgeniy Vorobeychik, Murat Kantarcioglu, and Bradley A. Malin. "Sociotechnical safeguards for genomic data privacy." *Nature Reviews Genetics* 23, no. 7 (2022): 429–445.

Wiseman, Yair, and Ilan Grinberg. "The trolley problem version of autonomous vehicles." *Open Transportation Journal* 12, no. 1 (2018).

Yazdanpanah, Vahid, Enrico H. Gerding, Sebastian Stein, Mehdi Dastani, Catholijn M. Jonker, Timothy J. Norman, and Sarvapali D. Ramchurn. "Reasoning about responsibility in autonomous systems: Challenges and opportunities." *AI & SOCIETY* 38, no. 4 (2023): 1453–1464.

For Product Safety Concerns and Information please contact our EU
representative GPSR@taylorandfrancis.com
Taylor & Francis Verlag GmbH, Kaufingerstraße 24, 80331 München, Germany